Our Insect Friends
and Foes

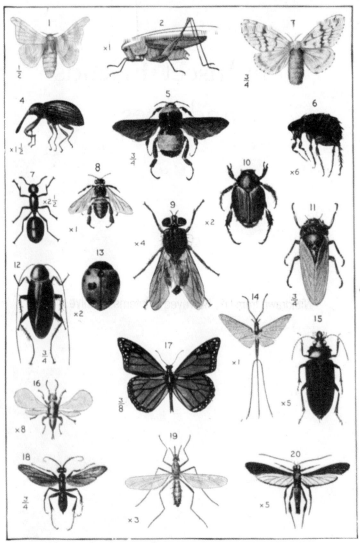

OUR INSECT FRIENDS AND FOES

1. Silkworm Moth. 2. Grasshopper. 3. Gipsy Moth. 4. Boll Weevil.
5. Bumblebee. 6. Flea. 7. Ant. 8. Honey Bee. 9. House Fly.
10. Japanese Beetle. 11. Cicada. 12. Cockroach. 13. Ladybird Beetle.
14. May Fly. 15. Boring Beetle. 16. Fig Insect. 17. Milkweed
Butterfly. 18. Hunting Wasp. 19. Mosquito. 20. Peach Moth.

Our Insect Friends and Foes

By

WILLIAM ATHERTON DuPUY

Introduction by
LELAND O. HOWARD

DOVER PUBLICATIONS, INC., NEW YORK

Standard Book Number: 486-22270-5
Library of Congress Catalog Card Number: 69-19466

Manufactured in the United States of America
Dover Publications, Inc.
180 Varick Street
New York, N. Y. 10014

PREFACE

THIS is not a book on entomology, but a travelog of insect land.

It is not intended to add to the knowledge of the scientist, but to show the general reader the vastly important relationship which exists between insects and human beings.

Its purpose is not to set forth weighty information in an impressive manner, but simple facts in such a way that the story of them will be easy to read.

It is not written for the occasional student who delights in dry tomes, but for the multitude whom it would help to discover that there is romance among these small neighbors that is as enthralling as that encountered by sailors shipwrecked on outlandish, far-away islands.

The language of science is not the language of the street. It must be technical, must use terms understood only by the initiated. These terms are commonplaces to the scientist. He uses them unconsciously much as each of us uses the words peculiar to the calling which he follows. The scientist cannot be stopped in what he is doing that the milkman may be made to understand. Getting the message to the milkman is somebody else's task, is the work for a go-between, an interpreter.

Thus does it happen that I, a man whose profession is not science, but journalism, have been thrust into the breech. I am to try to make Entomology understandable to and interesting to all men.

iii

Yet the book is to be scientifically accurate. Great care has been taken to make it so. It has come into being from the beginning under the watchful eye of one of the ranking entomologists of the generation. Dr. Leland O. Howard has been its godfather.

For nearly half a century Dr. Howard has been associated with the Bureau of Entomology of the United States Government, the largest bureau of its kind on earth. He has been with it from the time it started. For thirty years he has been its chief. Practical Entomology is a creature of the present generation, and this man has, perhaps, contributed more than any other to its development.

Dr. Howard has guided me in the preparation of this book, has lent all the facilities of the Bureau in massing my material. He has read the manuscript, corrected it, and submitted certain chapters for further reading and correction to men who specialize on particular insects.

Rarely has a book interpreting science been produced under such favorable auspices. Because of the unusual coöperation I have received I present it with confidence in its merit.

WILLIAM ATHERTON DUPUY

CONTENTS

ILLUSTRATIONS

INTRODUCTION

THE world is just awakening to the importance of the study of insects.

Most people are indifferent to them because of what appears to be their insignificance, and those persons who collect them and study them have been considered until recently as triflers or weaklings.

In poetry, in stories, and from the stage the entomologist has been ridiculed.

But within the last forty years there has come a change. It has become plain that of all the creatures that live on this earth the insects are the most powerful rivals of and enemies of man.

They outnumber all of the other kinds of animal life; they outnumber all flowering plants; they have inhabited the earth for more than fifty million years; they have accommodated themselves to all sorts of conditions; they have passed through cataclysm after cataclysm which swept away or seriously modified other forms of life.

In these fifty millions of years their structure has adapted itself to an infinite number of different modes of life, to an infinite number of different kinds of food; their structure, their physiology, their habits, the rapidity with which they multiply, all fit them for existence long after the newly-born human species shall have passed away.

They carry disease to man; they daily threaten his

life; they shorten his food supplies, both in his crops while they are growing and in such supplies after they are harvested and stored; they injure his meat animals; they destroy his clothing; they mine his houses; they disturb his sleep; they injure him in countless ways.

It is perfectly true, as Mr. Du Puy has pointed out in one or two of the chapters in this fascinating book, that in the long-time evolution of this enormous number of different kinds of insects and in their gradual adaptation to all kinds of food—some of them almost impossible to human imagination—many kinds have come to feed on other insects; and so numerous are these beneficial insects that they are man's greatest aid in the natural control of the crop-feeding species. It is almost safe to say, in fact, that were it not for this fighting among the different kinds of insects they would almost overwhelm the human race.

Appreciation of all these things is slowly coming, and this appreciation must become general, since all of us must back the scientific man in his study of everything about insects in order that we may secure eventual control. And there must be an enormously greater number of these scientific men to carry on the study. At the present time the schools and the colleges are hardly alive to the importance of the subject. The critical nature of the situation is only just being realized.

It is for this reason that I greet this book with joy. Not only must the scientific man know all about insects, but every one must know as much about them as he can be induced to learn. And so, when my friend, Mr. Du Puy, brought me these chapters one after the other,

I realized that here was a most important opportunity for me to help in a new way in the great and important work.

The stories he has written are wonderfully done.   No one has ever done them quite so well before.   I wish that I could command his services as a writer about insects for the rest of his life.

My own part in the book has been simply to see that what he has written is scientifically correct, and in this I have had the enthusiastic help of several of my expert colleagues.

L. O. HOWARD

Our Insect Friends
and Foes

CHAPTER I

THE FIG INSECT

IF anyone should ask you who put the box of dried figs in the fruit store window you would be likely to guess that it was the man who runs the shop or the boy who brought them on the wagon.

But if you should ask the same question of a certain wise man of my acquaintance who works for Uncle Sam and who knows as much about figs as anybody in all the world he would give you a quite different answer.

He would say that the figs were put there by a strange little insect not bigger than the gnat that gets into your eye in the summer time and makes it smart.

And this being such an odd statement for him to make you would insist on the story, which, when told, would lead a long way toward the claim that there is a romance back of every one of the commonplace things round about us.

These dried figs, it seems, are called Smyrna figs, which name alone carries the imagination far afield, and they come from California. In this country we eat as many of them every year as could be hauled by a thousand teams hitched to a thousand wagons—a string of them two miles long. Yet every fig on every wagon has had the personal attention of one of these small insects. One of them has died for it. Otherwise it would not have been on the wagon.

The fig, it seems, was one of the first plants in all the world to be cultivated by man. This was due largely to the ease with which it could be grown. In a warm climate one had only to cut off a branch and put the end of it in the ground and properly water it to produce a fig tree. Despite this it did not exist in America until it was brought over from its natural home in the Near East. It belongs over at the further end of the Mediterranean Sea, ten thousand miles away, in the region inhabited by the people who wrote the Bible. Smyrna, the city in Asia Minor just around the corner from the Holy Land, the city over which the Greeks and the Turks were fighting and which was burned not so long ago, is mother of the fig which takes its name. Smyrna boasted an ancient civilization before the time of Christ. It is believed that it was there in a cave on the River Meles that Homer wrote his poems, thus becoming the father of literature. And when men first began to record their thoughts, they spoke of their earliest friends of the garden—the vine and the fig tree.

When California was settled, travelers compared its climate with that of the countries that border the Mediterranean, and found that it was very similar. Some one thought of the figs of Smyrna, so highly prized for their excellence, and wondered if they would grow in California. There were already fig trees in California and elsewhere in the southern part of the United States, but none of this Smyrna variety which has peculiar qualities that make it better to eat than any other fig that grows.

So, back in the eighteen eighties, a newspaper in San Francisco sent to Smyrna for 14,000 fig cuttings and gave them to its readers. They were planted throughout the warm portions of the state and grew well and rapidly.

When these trees came to be four years old they should have borne fruit. In fact, each summer, many little figs started forth on their branches, developed until they were about half grown, then dried up and fell off. In all California not a single fig of this kind stayed on the tree until it was ripe.

Years passed. The fig trees grew very large. Their branches spread wide. Millions of little figs came out every summer, grew cheerfully for a while, then died and dropped off in the dust. The orchards yielded no fruit. They were complete failures. After all the trouble that had been taken to develop them their owners began going out with axes and chopping them down.

All through the eighteen nineties in California there existed the mystery of why the Smyrna fig trees bore no fruit. The best experts that could be found studied the problem in vain. In those days the Department of Agriculture in Washington was not very large and had very few scientific men on its staff. It is the policy of the government, however, to give practical aid to its citizens when they are in trouble. This riddle of the fig trees became so great that the government at Washington finally took it up and sent an expert all the way to Smyrna to find out what they did over there to make the little figs stick on until they were big figs.

There, among the strange peoples of Meander Valley where the figs grow best, this expert watched them cultivating and harvesting their crops. In the round of seasons there was one queer thing that he saw them doing, a thing that he had heard spoken of as a means of keeping away the evil spirits that might interfere with the harvests.

Outside the cultivated orchards there were what appeared to be wild fig trees. They were called *caprifigs*. On these trees there grew a dwarfed fruit never used at all as a food. At certain times of the year, however, the natives went to these trees, gathered the tiny dwarf figs, and strung them on pieces of fiber. Eight or ten of them would be tied into a loop. Then the natives would go

STRINGING THE CAPRIFIGS WITH INSECTS IN THEM.

through the orchards and hang one of these loops on every cultivated fig tree.

This American watched these caprifigs very carefully for there was a hint in the books he had studied of what might here be taking place. Sure enough, in a day or two, there was seen to come out of the "eye" of these wild figs, the opening at the end of them, a certain tiny insect, in fact, many tiny insects. He broke open one of the caprifigs and found hundreds of insects inside of it. They were strange little creatures, evidently belonging to the wasp family, and about as big as a mosquito with its legs trimmed down. It began to look as though they might be among that group of little insect animals which busies itself in doing certain tasks that are very helpful to man. Our scientist watched these little wasps to see what they did after they came out of the wild fig.

They flew busily about, he observed, evidently looking for something, the finding of which they regarded as most urgent and important. When they discovered the object of their search they went immediately to work upon it. That object was none other than the young fig on the cultivated tree. Their point of attack was the opening in the end. There they gnawed savagely and crowded excitedly to force an entrance. Eventually they pushed with difficulty through to the inside of the fig. So small was the opening and so hurried the entrance that the visitor often rubbed its wings off getting in. Presently it would reappear at the doorway, would scramble out, would fall wingless to the ground and die. It was the end of the trail for this excited little wasp.

The American scientist knew that insects, bees, for instance, often play a very important part in the growing of many kinds of plants. The bee goes about in search of honey, visiting one flower after another. In all these flowers there is what is known as pollen. It is a fine, dust-like substance. It is what you get on your chin when you play "Buttercup, buttercup." This pollen gets on the feet and legs of the bee. When that bee flies away from one flower to another it carries some of that pollen with it. When it walks over the threshold of the next flower it wipes its feet and leaves some of the pollen.

Every flower is in great need of pollen to fertilize it, but it must be pollen from another flower that is its natural mate. If the flower entered is the sort of flower that needs some of the sort of pollen that the bee has on its feet to fertilize it, it seizes these particles and makes use of them. Unless it gets the sort of pollen it needs, it will wither up and fall off without making any seed. If it gets the pollen it needs, it will ripen seed that may be planted and that will grow into other generations of its kind. Many plants are entirely dependent on these honey-hunting bees to get the pollen that means life or death to them. So does plant life come to be dependent upon insect life.

A FERTILIZED AND AN
UNFERTILIZED FIG.

Another odd thing that this American scientist knew was the fact that a fig was, in reality, a flower turned inside out. The petals of this flower grow inside a sort of hollow bowl. One must break the bowl open to see the flower. You can still see the dried flower in the dried fig. The fig insect must crawl into the "eye" of the fig to deposit the pollen which fertilizes it. He later found out another rather tragic fact about this little insect. When it came out of the caprifig it was looking for a nest, for a place in which to lay its eggs that baby fig insects might be born. It found its way into the fruit fig expecting to find a proper nest. Inside, however, he discovered that the place would not do. The furnishing was not right for an insect nursery. It was different from that inside the caprifig which had the proper arrangement for raising a fig insect family. So, after going all about and distributing pollen, it came out again, maimed and wingless. It could only drop to the ground and die, leaving the raising of young to carry on the race to a lucky insect that found a caprifig home instead of making the mistake of entering the fruit fig.

Still, if part of these insects did not make this mistake, the fig family would have run out long ago as none of the fruit would have been developed that produced seed. Nowadays the fig tree can be grown from cuttings and could get along without seed because man takes a hand. But in nature the ripening of seed was necessary to its continued existence. The caprifig as well as the fruit fig comes from these seeds, so, if no mistakes were made, a time would come when there were no caprifigs as homes for fig insects, as places for them to make their nests.

The very existence of the Smyrna fig tree depends on this insect. It serves but one purpose in the world, and that purpose is to carry pollen from the caprifig to the fruit fig. It, in turn, depends for its existence upon the caprifig, for there is no place else where it lives or can live.

This scientist in far-away Asia Minor drew his conclusions. The Smyrna fig tree of California, he argued, was not bearing fruit because it was not fertilized. Ordinary fig trees might yield ordinary fruit without fertilization. They did, in fact, but it was poor fruit and the seeds inside of it would not grow. This aristocrat of the fig family would refuse to bear fruit unless that fruit could be properly fertilized to make it of first class quality.

This tiny insect of Asia Minor, the like of which did not exist in America, was the thing needed. The solution of the riddle of the California orchards lay in getting these tiny wasps to that state, and there making homes for them in which they would grow and multiply.

But this was not as easy as it seemed. In fact it was such a hard job that it took as great an agency as the United States Government four years to accomplish it.

The fruit-bearing fig and the caprifig were both growing in California. The province of the caprifig tree, as I have said, is to make a home for the fig insect and to furnish it with pollen when it starts out. Its fruit is the nesting place of continuing generations of fig insects. That it may keep these tiny creatures going it must produce dwarf figs for them throughout the year. The fruit figs ripen but once a year—in the summer time. Small numbers of dwarf figs, however, are to be found clinging to the bare branches even in the winter time. In them generation after generation of fig insects follow one another. If there were a single two weeks in a single year when there were not dwarf figs available as homes for these insects, they would cease to exist on the face of the earth. All the Smyrna fig trees in the world would become barren as were those of California, and this delicious fruit would be lost to mankind.

The difficulty in getting the fig insect to the United States lay in the fact that there was not time enough between one generation and another. If dwarf figs were taken at the time when the eggs for a new generation were fresh laid in them and were started for America, the ship bearing them, under normal conditions, would not have got half way across the Atlantic before the insects had hatched and were ready to emerge. What they would need then was a new crop of dwarf figs in which to lay new eggs for the new generation. Not having them, they would die of old age in two or three days, childless, and there the enterprise would end.

Various schemes were resorted to and failed. Finally, one of the government's experts, knowing that the generations followed each other more slowly in winter than in summer, gathered some of these wild figs in northern Africa at this end of the Mediterranean which is nearest America. He wrapped them in tinfoil to keep them from drying out and kept them cool to hold back their development. Then he sent them in the greatest haste across the Atlantic and to California.

They got there on time. The fig insects were still alive.

THESE CAPRIFIGS ARE HUNG IN THE TREES.

They promptly came out and laid their eggs in caprifigs ready prepared for them. In a few weeks there were thousands and soon millions of them. The caprifigs with the insects inside them were taken into the orchards when the next fruiting season came around. These little creatures promptly swarmed out. They crawled into the "eyes" of the half-grown figs in the fruit-bearing trees round about, carrying the pollen of the caprifig with them. They came out and died on the threshold as is their wont.

And the miracle had taken place. Those figs, instead of falling off, as had been the custom of their fellows, stuck to their branches and began to swell and strut with a prodigious growth. They had been reprieved. The sentence of death was no longer their lot. With fertilization the flowers within them ripened into seed. About those seed were deposited sacks of sugar as food for them if they should ever be put into the ground for the purpose of growing new plants. This sugar gave the richness and flavor to the Smyrna fig as it lies there in the fruit store window awaiting your purchase.

Today, near each of the thousands of fig orchards in California, are caprifig trees. In the dwarf figs that they produce are bred uncounted millions of these little Mediterranean insects. When the season is just right, the owners of these orchards gather up these dwarf figs, full of these insects ready to emerge, and distribute them through the orchards. These Americans are more practical than the people of the East from whom they got the secret of fig fertilization. They do not take the trouble to string them on threads. Instead, they place a little wire basket in each tree, there to remain year after year. Then, at the proper season, they go through the orchards and put a handful of dwarf figs in every basket. These little insects emerge as is their custom, and the miracle is repeated.

There contributes to this miracle a little insect which gives up its life, a scientist who ferreted out this strange secret of nature, and a benevolent government that takes infinite pains to render service to its citizens to the end that they may live more happily and in the enjoyment of always more and more of the fruits of the earth.

QUESTIONS

1. What made the newspaper men of California start the "Plant-A-Fig-Tree" movement? Would you call the undertaking a foolish one? Before making your decision, compare the climate of California and of Asia Minor.

2. What prompted the government at Washington to send an expert to Smyrna to study fig growing?

3. Prepare an interesting story on "How One American Scientist Used His Eyes in a Smyrna Fig Orchard." In telling your story, keep the following big topics in mind:
    (a) Comparison of cultivated or fruit figs with wild or caprifigs.
    (b) Superstitious customs of the natives.
    (c) Meaning of the insect's visit to the caprifigs.
    (d) The insect's work.
    (e) Dependence of fruit fig on the wasp insect.

4. What does the author mean when he calls the fig "a flower turned inside out"? Examine a dried fig, find the eye, and open the bowl of the fig.

5. Explain the following statements:
    (a) "Every Smyrna fig has had the personal attention of one of these small insects. One of them has died for it. The insect's loss is our gain."
    (b) "The dependence of one upon the other is just as essential in the fruit or plant world as it is in the animal world."

6. Why does Mother Nature put so much sugar about the tiny seed of the fig?

7. (a) Why did it take the United States Government four years successfully to bring over the fig insect to fertilize the Smyrna fig?
    (b) If this experiment had not been successfully made figs would cost the American citizen much more. Why?

8. Describe a visit to a California fig orchard. What evidence of American practical methods do you see?

VOLUNTEER WORK

9. (a) Why should we be proud of Uncle Sam's workers? Give several reasons why a scientist should use his eyes.
    (b) The early writers often mentioned the fig tree in their books. See how many references to the fig you can find in the Bible. Read the parables of the fig tree in St. Mark and St. John.

Chapter II

THE BUMBLEBEE

IF all the bumblebees in the world were killed forthwith, there would be no more red clover.

When one makes a wish on a load of hay, he is not likely to know that a buzzing bee with a long tongue may have had as much to do with its making as did the man with the pitchfork.

Members of the insect world have a peculiar importance that is not generally understood in many of the activities of man. They save the peach crops for him, for instance, and may destroy the wheat crops. Insects are good and bad—mostly bad.

The bumblebee is a helpful insect. There would be no red clover hay if it were not for this creature—the droning, clumsy, sociable, care-free wanderer of the meadows —the bumblebee. There would be no red clover hay, in fact, if it were not for that last sixteenth of an inch on the end of the tongue of the bumblebee. Mere honey bees can help with the white clover, but in the fields of red clover the bumblebees hold a labor monopoly that cannot be broken. The bumblebee is the world's champion haymaker, and his success is due to the fact that he has a longer tongue than the other members of the bee family.

No less a person than Uncle Sam himself had a very

peculiar experience in haymaking and solved an industrial mystery through his knowledge of bumblebees. After the Spanish-American War the United States found that it was in possession of a group of tropical islands sitting on the equator pretty well around toward the other side of the world. The Philippines were a pleasant land inhabited by brown men who had never been given much of a chance, so Uncle Sam wanted to lend a hand wherever he could toward making them happy and prosperous. Much to his surprise he found that in those islands there was no clover, and clover was a big help to both Europeans and Americans. Forthwith he decided to present this pleasant crop to the Filipinos. He took clover seed over, planted it in properly selected fields, and it grew vigorously.

But if one returned to these clover fields a year or two later he would find that there was no more clover. It had mysteriously disappeared. The Philippines were as bare of clover as they had been in the beginning. The clover had failed to reproduce itself. The crop that had grown so vigorously had made no seed. Without seed there could be no succeeding crops of clover. Something was wrong.

The Department of Agriculture was asked to solve the mystery. It was wise in the habits and needs of plants. It knew much, for instance, of the clover blossom. All blossoms must be fertilized or they will not make seed. They must have dust-like pollen placed in them. This pollen must come from other flowers of their kind. They cannot go after it, so they depend on some insect friend to bring it to them.

It is because they want bees and butterflies to bring them pollen that the flowers of most plants produce a sweet morsel of honey. They know that bees like honey, so each flower deposits a bit of it deep down in its heart as a reward to the bee that will bring the pollen it needs.

The honey drop which the clover blossom deposits to pay for this favor is at the bottom of a very deep cup. None of the butterflies and very few bees have tongues long enough to reach in and get it. The bumblebee, however, the biggest of all the bees, is just fitted for the

THE HONEY-CARRYING LEGS OF THE BUMBLEBEE.

task. It has a tongue of the required length, and it has a fuzzy head on which are sure to be sticking many particles of pollen from flowers it has formerly visited. As it gathers in the honey money that is its pay, it shakes down a few particles of pollen to the very bottom of the flower. It has earned its wage. This flower will make seed that new generations of clover may follow.

The government scientists knew all this. Their books of insects, their entomologies, told them that insects were rated as a "class" in the arrangement of the animal kingdom, and that, in this class there was an "order" known as the Hymenoptera. The Hymenoptera were insects with membrane wings (hymen in Greek means membrane and pteron means wing). The principal members of this order of insects were the three super families, the bees, the wasps, and the ants. In the bee family the bumblebee is a sub-family. This sub-family did the work of the clover fields.

Now, when clover was planted in the Philippines, there were plenty of bees, but no bumblebees. The Philippine bees had short tongues. None of them could reach to the bottom of the clover blossom. They could not get at its honey. They knew this instinctively, so they did not even try. So no bees came to the clover fields, and the clover made no seed.

The problem of the government in making clover grow in the Philippines was to establish the bumblebee over there so that it could perform its customary tasks. Agriculture often depends on other crops than plant crops. A crop of bumblebees must be raised in the Philippines.

Here again the success of an undertaking depended on an odd sort of knowledge. How, these scientists asked themselves, can we best take bumblebees to the Philippines, they being untamed creatures with a busy business end when aroused?

But they knew the life cycle of the bumblebee—the way it lived from the cradle to the grave. They knew that nearly all the bumblebees, the workers, the nurses, the males, died in the autumn. Only one type, the queens, lived through the winter. They might succeed in doing this by crawling up into the middle of a shock of corn in the field. A safer place for them, however, would be a rotten stump or log. They might crawl far back into it, through tiny crevices, thus making it impossible for their arch enemy, the mouse, to get at them and make a winter meal off them.

In the rotten log they hibernate. They curl up and become as though dead. Through the long winter they do not even breathe. They know, and nature knows, that they must go all winter without food, and, since any movement would use up their reserve strength, they remain absolutely dormant. But within them the spark of life glows dimly. When the warm days of the spring come they thaw out, stretch themselves, creep from their hiding places, and are away again to the meadows as good as new.

These scientists went out that winter and split open various rotten logs until they had collected a considerable supply of these hibernating bumblebees. They knew that they did not dare take them into heated houses or they would thaw out and get busy. Yet to travel to the Philippines they had to travel by heated trains and boats. They even had to travel into a part of the world where it is always summer. Their transportation was a problem.

The solution was to refrigerate them. They must not get the idea that springtime had come. They should be transported half around the world inside of ice boxes, where it seemed to be winter. They were so transported. Then, having arrived in the Philippines, they were allowed to thaw out. Having done so, each queen set out on her own responsibility to plant a bumblebee colony. They seem to have been successful, for now bumblebees are plentiful in the islands, and clover grows in abundance.

Thus did the introduction of the bumblebees make clover possible to the Filipino. It has likewise made it possible to those colonists of Great Britain who settled in far away Australia, New Zealand, and South Africa. It was introduced by the British into those regions just as it was introduced by the Americans into the Philippines. The English sometimes say that the bumblebee won the Boer War. They would not have been able to defeat the Boers had they not been able to feed their cavalry horses. They could not have fed these horses had clover not been established in South Africa. Clover could not have been grown there but for the coming of the bumblebee. So there you are.

These scientists, these entomologists of the Department of Agriculture, know exactly what each of these queen bumblebees turned loose in the Philippines did. Each looked about for a suitable place to start a nest. Often a deserted nest of a field mouse is selected, or a clump of moss or grass is collected. With this as a protection the queen bumblebee begins her visits to the clover blossoms. There she collects honey and pollen which she mixes together and forms

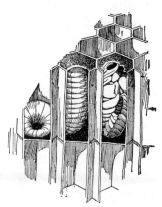

GRUBS OF THE BUMBLEBEE IN CELLS.

into a primary cell. In this she lays an egg. Her family is started. She works busily adding one cell after another to her home, laying an egg in each, gathering the honey, feeding the babies, doing all the housework. These tasks are later divided among the members of the colony, but at first there is only the queen mother to do everything. The eggs hatch in two or three days and develop into worm-like larvæ or grubs. These eat and grow until they get to be an inch long and as big as your finger and finally arrive at what is known as the pupa stage. In this stage they become hard, lump-like bodies and lie still while changing from grubs to bumblebees. When the pupa becomes a bumblebee it eats the top out of its cell and comes out.

As soon as the young bees appear, they take up their share of work. They first go into the field and gather honey. As the colony grows larger, there is a division of labor and some become houseworkers, some nurses, and some honey gatherers. The big fellows go for the honey and the little ones do the housework. The middle sized ones usually serve as nurses, carefully mix the honey and the pollen, and feed the grubs through tiny holes made by other worker specialists. The queen no longer leaves the nest. She does not dare do so because other members of her brood, despite their willingness to work for the young ones, have an irresistible fondness for the eggs the queen lays and, if they can catch the mother for a moment off her guard, they will eat them. So the queen stays at home and watches her eggs. Presently young queens appear and themselves lay eggs. Oddly, the mother who has all the time fought so vigilantly for the protection of her own eggs and who still does so, now turns vandal and attempts to steal those of her own daughters. No bumblebee, however, resists the queen who stands guard. They slink away when they are caught stealing eggs. They seem to feel the weakness of being in the wrong, while the defending mother, with right on her side, has the strength of a mighty host.

When the young are out of their cells, the moppers-up go in and give them a thorough cleaning. After that the cells are used as storehouses for extra food. The bumblebee does not store much food, for it knows that it is not going to spend the winter here. It merely keeps a supply sufficient to carry the colony through a rainy spell of two or three days or any such emergency. In a bumblebee nest one may find honey from the amount of a spoonful to a cupful, but rarely more.

This honey, however, has been the basis for many a battle royal between farmer boys and bumblebees. There is difference in report as to its sweetness. Boyish recollection is likely to give it an incomparable sweetness which it is not found to possess when grown men go out later to test it. But the fight for it is great sport when the attackers, armed with handy paddles, stir up the nest, an act which arouses intense anger on the part of the inmates. There are no slackers in the bumblebee colony. They go into action to the last individual. They sally forth and begin to circle around like so many airplanes attempting to locate the enemy. Having found him, they proceed to business. They know their business and strike for the enemy's most sensitive point, the eyes. An all-wise nature, however, gave boys an instinct for protecting their eyes. So quickly and instinctively can the head be moved that the dart rarely hits the eye itself, but is likely to land in the face near it. A bumblebee rarely strikes except in the face. His stinger is half an inch long. If it is examined under a microscope it will be found to consist of two parallel shafts bound tightly together. There are notches on the insides of these that fit into each other. When the stinger is inserted the two halves of it work quickly on each other with a jack-screw effect, thus pushing the point deeper and deeper. When it is in as far as it will go the bee squeezes a little bulb which injects bits of poison that it keeps on hand for the purpose. It is this poison that adds to the pain of the sting and that causes most of the swelling.

But the attacking boys try to strike the circling bumble-bees with their paddles before the bumblebees can drive home their stingers. It is a thrilling fight with possibilities of success on either side.

The farmer boys of Kansas accidently discovered a method of attack on a bumblebee's nest by which, through strategy, much completer results could be obtained than through open attack. In Kansas it was the custom to take drinking-water to the men working in the hay fields in large jugs. This task naturally fell to the boys. It is probably true that at some time a young boy with a jug stopped in his work as water carrier to attack a bumble-bee nest and by so doing made a discovery. At any rate these Kansas boys have a very effective method of their own. It is this:

They set a jug, partly filled with water, with the stopper out, near the bumblebee nest. Then they stir up the bees and get entirely out of their way. A very peculiar thing happens. The bees come out and begin circling angrily about and buzzing noisily. The only strange thing they see is the water jug. They approach it threaten-ingly. Their buzzing awakens an answering buzz, an echo down the neck of the jug. The mad bees regard this as the challenge of an enemy. They fly at the mouth of the jug and tumble inside. There they beat about and increase the noise. This centers the attention of the other bumblebees. One by one they strike at the neck of the jug and one by one they plunge inside. Soon they are all trapped. To make sure that the victory is complete the boys may return and stir up the nest a second time and trap what fighters remain. Then they help themselves to the honey with safety. After it is all over, the water, containing the half-drowned bumblebees, may be poured out and the fighters will get dry and recover. It is a good idea to give the bumblebees a chance to live just as it is a bad idea to go out of your way to destroy their nests, for they are good friends of man and do for him a thing no other creature in the world could do.

There are many interesting facts about the lives of bumblebees that are quite well known. We know, for instance, that they work night and day when the nights are clear and bright. When the night is dark, they rest and sleep. On such occasions, however, they place a sentry on patrol and this sentry goes busily about all through the night. The same bee always serves as sentry. If that bee is removed another is appointed and performs all the sentry duty. How this appointment is made in the bumblebee colony, whether the sentry is elected according to the American plan of choosing individuals for public posts, or whether the post is hereditary, as in the older eastern forms of government, is not known. The sentinel bee keeps its eye open for the first appearance of dawn, and when it comes this bee drums noisily with its wings and wakes the entire colony.

Another odd thing about the bumblebee household is the fact that there is another bee, known as the guest bee, a cousin of the bumblebee, resembling it very much in appearance, that crowds itself in and makes its home with its relatives. This self-invited guest, while like its host in appearance, is very unlike it in nature. It is, for example, very lazy. It has no house of its own. It does no work. It does not even have pollen sacks on its hind legs as do the worker bees. It can be told from the bumblebees by this difference. It eats the honey that is brought in by its hosts. It lays its eggs in the cells made by the bumblebees and allows the busy workers to feed its young. When it first arrives the bumblebees do not seem to receive it very politely or graciously, but it is an agreeable and sociable creature and is soon accepted.

How, you may well ask, do the entomologists learn all these secrets of bumblebee life? Simple enough. You, yourself, may establish a bumblebee colony and observe as much. All you need to do is to follow a few simple directions. Get a deep cigar box, fit it with a glass cover, and make a hole in one end big enough for bumblebees to go through. Then go out into the fields and find a bumblebee nest, marking its location so carefully that you can go back to it in the dark. Return to it at night with your cigar box and a bottle of chloroform. Pour the chloroform on the nest and listen carefully until all the noise within it ceases. Take up the nest and install it in your cigar box. Return home and arrange a place for it on your window sill so that the glass-covered nest may be seen from inside the room, but with the hole through which the bumblebees come and go on the outside. The colony will work on your windowsill through the season and you may observe all that takes place.

THE BUMBLEBEE

QUESTIONS

1. (a) How did America try to help the Filipinos in getting better returns from their land?
   (b) What "industrial mystery" resulted from this enterprise?
2. We all know that bees and flowers belong together. The following questions will help you in summing up the important parts of the story:
   (a) Are bees the friends of flowers? Prove your statement.
   (b) Mother Nature generally makes provision for what she wants. How do flowers attract the bees?
   (c) Why is the bumblebee the world's champion haymaker rather than the honey bee?
   (d) What loss would result if all the bumblebees throughout the world were forthwith killed?
3. If you sit down near a patch of red clover on a sunny day in a short time you will be sure to see a bumblebee light on the clover. In describing Miss Bombus note the following:
   (a) Size.
   (b) Fuzzy head.
   (c) Pick a head of clover to pieces and find the honey.
   (d) Ways in which it differs from the honey bee.
   (e) How the flower rewards the bee for its visit.
   Do you think the bumblebee "has earned its wage"? Why?
4. Bees, in their daily life, obey laws that civilized man often neglects. Find statements in the story that illustrate the bee's observance of the following laws:
   (a) Order.
   (b) Division of labor.
   (c) Cleanliness.
   (d) Honesty and fair play.
   (e) Care for the young.
   (f) Thrift and conservation.
   (g) If you were a member of a bumblebee colony, what position would you like to occupy? Why?
5. (a) Would it have been at all possible for the Spanish buccaneers of the fifteenth and sixteenth centuries to solve the transportation problem and get the bumblebees over to the Philippines "on time"? What modern invention helped Uncle Sam to carry out this important experiment?
   (b) How was red clover introduced into the British Colonies? What

do the English mean when they sometimes say "the bumblebee won the Boer War"?

6. What do you think of the strategy resorted to by the farmer boys in Kansas? A young farmer lad who had participated in many such pranks himself read this article. When he put the book down he said, "Aren't farmer boys foolish in hurting their good friends?" Do you think this lad will appreciate the work of the bumblebee in the future?

7. Do you agree with what Hornaday says about wild life? "The wild life of today is not wholly ours to dispose of as we please. It has been given to us in trust. We must account for it to those who come after us and audit our records."

8. If you have ever carried on any experiments with bees, tell your experiences to the class.

You may have been on a farm during your vacation. Were the men on horses stung when the hay was being cut? What does this prove?

CHAPTER III

THE GRASSHOPPER

INTO the towns of Tripoli, in North Africa, there come the solemn processions of mules heavily laden with bags well filled with nourishing food for the people, now about to be put on sale in the market place. These bags contain dried grasshoppers.

The rain doctor of the Hottentots, at the other end of a great continent, is busy with his sacred rites. Famine is upon his tribes and he is praying his Gods to send to them that blessing which, like a miracle, has often before come out of the desert. May his people be favored, he asks, with the coming of a swarm of grasshoppers that they may feast without end and grow fat and strong?

The Bedouin of the desert has made his camp. In a hole in the sand a red fire is burning. It has heated the earth all about. Now the brands are taken out and the hole is filled with a huge bagful of squirming grasshoppers. Hot sand is piled in upon them and the fire is rebuilt over them. So does he prepare for a rare feast.

Arabians in Bagdad dine deliciously on grasshoppers and pomegranates. About the Red Sea boiled grasshoppers are spread upon the roofs of houses to dry, then placed on ships to become articles of commerce and items on the bills-of-fare of distant cities. The Moor prefers them to pigeon. They are sold in Chinese towns as are roasted chestnuts in Philadelphia. Filipinos parch them in earthen pans. The Bushmen of Australia grind dried grasshoppers into meal from which they make cakes. Early chronicles of the West Indies record a traffic in grasshoppers. The Digger Indians in California indulge in grasshopper drives, bringing the insects together at a central point where a trap for them has been built. In fact, grasshoppers have for centuries been used as food by something like half the people of the world.

This insect of commerce is the short-horned grasshopper more generally spoken of all around the world as the locust, a big, meaty fellow often two inches long. In any meadow on a summer day, anywhere in the world, may be found both short-horned and long-horned grasshoppers. No insect is better known or more widespread.

The horns are, in reality, the feelers, or antennæ, handy organs with which it feels, hears, and smells. If a grasshopper's antennæ are shorter than its body it is a locust. If its horns are longer than its body then it is a meadow grasshopper with no special name.

The locust is still a grasshopper just as an Airedale terrier is still a dog. It is a special breed of grasshopper. It is the most important breed of grasshopper there is. It might also be said that it is one of the most important insects in the world. This is not merely because it is so widely used as food, but because it has so often played tragic parts in the affairs of man—has so often thrust famine, plague, death upon him. Millions of men in many lands, since time began, have gone to their death through starvation and plague caused by the locust. The locust has played no mean part in the history of the world.

The Bible tells of the plague of locusts in Egypt that "covered the face of the whole earth, so that the land was darkened, and they did eat every herb." History records many such visitations. When Carthage and Utica, on the north coast of Africa, were in their glory back in Roman times, grasshopper swarms, now and again swept to sea by the winds, drowned in uncounted millions, and washed ashore. Barriers of them along the beaches, reaching for hundreds of miles and four or five feet high, produced such stenches that plagues followed, and hundreds of thousands died.

Africa is the home of grasshoppers. Within its great waste areas they breed in stupendous quantities. Now and again a scourge of them sweeps down on Morocco, Tangier, or some other border state with resulting devastation and pestilence.

Occasionally they ride the winds and leap stupendous barriers, such, for instance, as the Mediterranean Sea. History shows these visitations, now to Italy, now to Spain, now to France. In 1744 all Europe was swept by these grasshoppers from Africa. They stayed for years and extended as far north as Sweden.

In 1825 a traveler reports journeying from Moscow to the Crimea, in Russia, and encountering such masses of locusts on the ground as to make his carriage drag as heavily as though the earth were freshly plowed; these swarms continued for four hundred miles.

A wanderer in India in 1811 encountered a swarm of grasshoppers which was forty miles across and three days in passing. In a single city in their track, Ahmedabad, half the population of 200,000 died as a result of this visitation. Wherever they passed there was no food left for man or beast.

The United States, no longer ago than 1876, felt the stress of the grasshopper invasion. Swarms of them arose from the waste lands where the Great Plains rise into the Rocky Mountains, laid out their courses to the east, and bore on unceasingly for days and weeks, flying distances as great as 1,000 miles. They reached those areas in Kansas and its neighbor states where fertile farms had been developed, and rich crops were pushing on toward harvest time. They alighted and ate every living blade of grass, every leaf and twig. They left the farmer nothing for the harvest. They left nothing upon which his cattle and horses might feed. They swept the country clean, leaving devastation in their wake. The loss because of this visit was placed at $200,000,000.

They disappeared, but there remained in the breast of the western wheat farmer that haunting fear that, maybe, the next year or the next the plague would return and ruin would stalk across the prairies.

Such is the record back of this king of insects, sunning himself so idly among the grasses on an August afternoon. Little indication does he give when one meets him casually of those conflicts of the past in which he has caused the death of hordes of these man-creatures with their vaunted superiority. Yet with such a past we might well pay him the compliment of curiosity. We might well inquire into the manner of creature this is that man has met so dramatically on his journey down through the ages.

Aside from its food value to man, aside from its tragic

contacts with him in the past, the grasshopper has yet
another claim to attention due to the fact that it is the
largest creature in the insect world.   What the elephant
is to the larger animals, what the ostrich is to the birds,
the grasshopper is to those six-legged creatures, known
as insects.   The short-horned locust, two inches long,
is a Gulliver to the insects; yet it is a small individual
compared to that giant of the South, the lubber grasshopper of Florida and Georgia.   This one is as big
as a bird, is three inches long and very heavy-bodied, so
heavy-bodied in fact that it gave up flying many generations ago and has so long
failed to fly that its wings
have become ridiculous
little frills on a broad back.

The grasshopper is a
cousin to that prowler in
the kitchen, the cockroach,
to that pious creature of
the leaves, the mantis or

GRASSHOPPER AT REST

devil's horse, to the cricket, and to others. These kinships in
the insect world are proved by the facts that these different
groups are alike in some important respect and are different
in that respect from any of the other insects.   The scientists
classify the grasshoppers and their relatives as *Orthoptera*,
which is a Greek word meaning "straight wings." All these
insects have a hard shell-like wing on top, of little use
in flying, and underneath it the really effective wing all
folded up in straight lines for all the world like my lady's
fan.   These are the only insects in all the world that fold
their wings this way.   They are the straight wings.

The grasshopper, its dressy half-sister, the katydid, and its sprightly half-brother, the cricket, differ from their other relatives in the fact that they are primarily jumpers and not runners. They are harnessed to a pair of hind legs that can be released at any moment, like a spring, and will hurl them great distances through the air. No other insect except the fleas are so fitted out. If a man could jump as high in proportion to his size as a grasshopper he could vault the courthouse and break every bone in his body coming down.

The hind legs of the grasshopper are likewise used for purposes of self-defense. The mule is by no means in the same class as a kicker with the grasshopper. The mule may kick another creature its own size and the only result is a bruise. When a grasshopper kicks a bumblebee, a thing it stands quite ready to do, that dull droner goes tumbling handsprings.

The grasshopper has yet another point of distinction setting it off from the other members of the insect world that is even more pronounced. With only one exception it is the only musician in all those millions of species that are called insects.

The impression is likely to get abroad that insects are a noisy breed. The bees, to be sure, make a humming noise when flying and the propellers of the mosquito may be heard if it gets close enough to the ear. Some flies also may drone a bit through the air holes in their sides. But the only insects in all the world that deliberately set out to make a noise is the grasshopper, sister katydid, brother cricket, and that one member of another order, the cicada. or harvest fly.

At that, grasshoppers are silent all through their youth—beat no tin pans until they are full grown. Even then the gift of noise-making is given only to the males and the mothers go to their graves without ever having broken the silence.

Grasshoppers do not sing. They have no voices. How could they have voices, being insects without even lungs? They are instrumental musicians, chiefly fiddlers. Their wing covers are their fiddles and their jumping legs are the bows. Since they do not have wings until they are grown, the youngsters, naturally, are silent.

Some grasshoppers play their simple tunes by drawing their jumping legs across these stiff wing covers. There is yet another manner of noise making—that of rubbing the back part of the wing cover against the fore part of the collapsible wing. Still cleverer musicians, as the katydids, raise both wing covers above the back and rub their edges together. They are excellent sounding boards and the result can sometimes be heard for a quarter of a mile. Then there is the clacking which the proud male accomplishes when flying.

This grasshopper was hatched from eggs laid in the ground. The female has a well-drilling instrument, and with this, in the late summer, she digs a hole half an inch deep. In it she lays as many as a hundred eggs. They stay there throughout the winter and in the spring they hatch out. The mother never sees her child and it must grow up an orphan and hustle for itself. The tiny grasshoppers look like their parents except that they have no wings. They do not become grubs, these pupa, in the course of their development, as do the bumblebees.

Like the young of all insects the chief mission of these baby grasshoppers seems to be eating. Like the young of even man they tend constantly to outgrow their clothes.

This need on the part of the young insect for changing its clothes develops a very remarkable thing about insects in which they are entirely different from any other creatures in all the world. The young insect grows inside of its skin, but the skin itself does not grow. In the

course of time it becomes too tight and, as the young insect refuses to stop growing, it must find another way out of the difficulty. What it does is to swell out its chest, hunch up its shoulders and burst its jacket down the back. After this is done the growing insect crawls out of it and sits there on its perch, a damp, skinless, flabby, and quite helpless creature.

Just how helpless is this insect which has just shed its skin cannot be well understood unless it is known that this skin is more than a covering for the body. It is likewise the skeleton which holds it in form. Where the higher animals wear their skeletons inside and cover them with their muscles, the insects wear their skeletons in the form of a sort of shell on the outside and attach their muscles to them on the inside. This skin, therefore, serves the double purpose of a covering to protect their soft bodies and the bones to which their muscles may be attached.

HE HAS JUST SHED HIS SKIN.

When the young grasshopper, therefore, outgrows his jacket—splits it down the back and crawls out of it— he is for the moment without covering or skeleton. He is a soft, flabby creature. His hopping legs, for instance, usually so stiff, are like so much twine. They have no stiffening at all. They are so limber, for example, that he could take them out of the hard shell of his old boots without even straightening out his legs. They could be pulled around the crook in his elbow like strings.

All the creatures of the insect world, during the time that they are growing, burst open and shed their old skins. The skins of different kinds of insects are to be found all about during their growing summer season. Each of these insects that has crawled helplessly out of its jacket has done the same thing. It has pulled itself together there by its old clothes, has struck an attitude, has assumed the shape into which it is to harden. Each has been a damp sort of creature, covered with a kind of gum. This gum is made of a substance known as chitin, which is a good deal like that from which the horns of cattle or the finger nails of human beings are made. When it comes in contact with the air, it immediately begins to harden. Very soon the insect finds itself with a new covering of this hardened chitin which furnishes it with a very durable suit of clothes as well as this outside skeleton.

Each time the grasshopper or any other insect makes any considerable degree of growth, it is necessary to burst out of the old skin and put on a new one. Most insects shed their skins four or five times in the course of their development. The last change of clothes comes at the time when the insect gets grown. When it comes out of this last suit of childish clothes, it usually appears wearing wings, a thing that it has never done before. It is now mature. The suit which it now wears will never be changed. An insect with wings never grows any more.

With maturity comes a lessening appetite. The grown grasshopper does not think so much of food. It is now midsummer, the insects' season of joy. The days are warm and long. All nature is rioting in abundance. It is frolic-time, mating-time. Wings have come to this monarch of the meadows. He is no longer a lowly creature that must crawl and hop along the ground. And with wings comes the power of music.

The music of the grasshopper is a love serenade. He usually plays to his sweetheart. She sits demurely near and gives little evidence of being impressed, although the antenna through which she listens must be a-tingle with the vibration of his song. The spirit of the meadows is that of cheery happiness. So does summer pass into fall.

The mother grasshopper possesses a discerning wisdom when she prepares for the little ones. Possibly she has lived all summer in the garden where the family vegetables grow. When she comes to plant her eggs, however, she leaves it. She seems to know that this garden will be plowed before her eggs can hatch and if they are turned under they will lie too deep in the ground ever to get out when the little ones are born.

She must find a place which will not be reached by the

plow. Along the edges of the garden, near the fence, there is hard ground, adapted to drilling holes, and here the plow never comes. Here she lays her eggs. It is from the nests of such wise mothers that the new generations of grasshoppers come. It is a knowledge of the advantages of plowing that is helping the farmer keep the pest down. The wise farmer in grasshopper country leaves no idle and unplowed land about.

This same necessity of waste land in which to lay their eggs is the explanation of the origin of hordes of migratory grasshoppers. Africa is the source of more injurious swarms of them than all the rest of the world. It is because there is in Africa so much waste land where their nests are undisturbed that so many of them are there produced.

The Rocky Mountain locust which attacked the western farmers back in the seventies came from the waste spaces where the Great Plains rise into the mountains. There were limitless undisturbed hatching grounds. From them rose myriads of insects, grown to maturity. Perhaps there were too many to find food at home. Perhaps there was within them the instinct for pioneering —for planting their flag in new territory. So they took to wing.

The thorax of this insect, its chest, is a sturdy block of strength. To it are tied the muscles that drive those legs in their business of leaping. To it are tied those wings that have such strength that they can push on and on without rest for days or even weeks.

But in this great American flight the result was peculiar. The Rocky Mountain grasshopper got into the great lowlands of the Mississippi where grew an unlimited food supply. When autumn came the mothers made no attempt to go all the way back to the mountains to lay their eggs. They found the best places they could and there made their nests. Their eggs followed the normal course and, in the spring, there was a new generation of grasshoppers. But they were not strong. The need of the vigor of mountain air was in them. They grew sickly. Few of them lived to maturity. This, it may be said, was a very fortunate fact from the standpoint of the farmer, who was by no means sure of the passing of the pest.

For a generation the farmers of the Mississippi Valley have been wondering if the grasshoppers would return— have been dreading the possibility of their coming back. This knowledge of how and where they breed throws some light on the chances of their doing so. All the time in the West more and more land is coming under the plow and so is being spoiled as breeding places for grasshoppers. As the uncultivated spaces grow less, so does the danger of the return of the grasshopper plague grow less. It is doubtful if they can ever again develop in sufficient numbers to repeat the exploit of 1876.

Here is an odd thing about grasshoppers at the seashore. In the sands along the beach there are many of them. They fly up as you come along, whiz noisily away, and alight. Look closely, however, and there they are in the sand, colored so exactly like it as to lose their identity.

Go back a bit from the beach where the grass is thick and obscures the ground, and there, also, are grasshoppers.

You scare them up as you wade through the grass. They buzz away and alight and are equally hard to see. The reason, you will find, is that these are green grasshoppers of just the hue of the grass on which they live.

If you put these green grasshoppers down on the beach sand or the gray grasshoppers out on the grass, both would be quite conspicuous. You could see them easily. So could their enemies, the birds. Both kinds of grasshoppers would be promptly eaten up.

In fact the green grasshoppers on the sand and the gray grasshoppers on the grass were all eaten up a hundred thousand years ago. But those that mimicked their surroundings survived and are still there.

The necessity for concealment runs all through grasshopper life, for they are the staff of life for many birds. Blackbirds, robins, crows, and larks feast upon them all through the summer. A butcher bird is never happier than when impaling a fat grasshopper on a thorn, there to tear it to pieces. To many birds, however, the grasshopper grows so large that it may not be swallowed. It can be eaten by them only when it is small, but at that time it is consumed in great numbers. Then toward the last it is the choice morsel of the huntsman turkey, growing fast against the coming of Thanksgiving, and the greatest grasshopper eater of them all. To avoid all of which the insect fades as much as it can into the background that may be its place of abode.

QUESTIONS

1. How can you tell a locust from an ordinary meadow grasshopper? What other relations of the locust do you know? Prove the relationship.

2. Your book cites many illustrations telling how the earth has groaned under its devastations.   Tell one of the following stories:
    (a) Locust plagues in Bible days—Egypt.
    (b) Locust plagues in Roman times—Carthage, Utica.
    (c) Locust plagues in modern times—Italy, Spain, France, Russia, India.
    (d) Locust plagues in the United States.

3. (a) Why is the locust the champion jumper?
    (b) Explain the following statements:
        The locust is the king of insects.
        The locust is the mule of the insect world.
        The locust is the musician among the insects.
        How is the "music" produced?
        Which member of the family obeys the old adage, "silence is golden"?

4. Before the grasshopper appears wearing wings, how many stages has it already passed through?   Describe the various stages of its growth, keeping the following points in mind.
    (a) Food.   Appetite.
    (b) Legs.
    (c) Changes taking place.
    (d) Thorax.   Wings.

5. (a) What takes the place of "bone" in the grasshopper to make the body so still?
    (b) Why do grasshoppers wear their skeletons outside?
    (c) Where do you find their muscles?

6. (a) In what kinds of places do grasshoppers flourish?   Why is Africa more the center for injurious swarms of locusts than other parts of the world?
    (b) What part of the United States affords a breeding place for the locust?
    (c) What can the farmer do to help diminish the dangers of plagues from locusts?
    (d) Give several reasons why the farmer should know something about science.

7. (a) Grasshoppers are the staff of life for many birds.   Illustrate.
    (b) How do grasshoppers protect themselves from inroads of the bird world?

VOLUNTEER WORK

8. (a) In the Bible we are told how John the Baptist lived while preaching in the wilderness.   What formed his chief diet?

  (b) Read of the plague of locusts in Exodus 10:15.

  (c) Grasshoppers have found many admirers among the poets. Read what Leigh Hunt, John Keats, James Whitcomb Riley have to say about grasshoppers.

CHAPTER IV

THE MILKWEED BUTTERFLY

NE of the most radiant, colorful, beautiful creatures in all the world is the milkweed butterfly—beautiful but odd, having strange qualities that you never would suspect. Who would ever believe, for instance, that they are as great travelers as the mallard duck that winters in the South and summers in the far North? Who would know that they have a sense of smell a hundred times as delicate as that of the trail-following bloodhound?

One must go a bit afield with the scientists to learn the strange tricks and abilities of these insects. Our learned friends will tell you that this milkweed butterfly, for example, despite its seeming frivolity and frailty, is one of the cleverest, most venturesome, most persistent of all those six-legged animals known as insects.

Before you hear the story of the milkweed butterfly, the scientist wants you to know that it belongs to that order of insects known as *Lepidoptera*, a word made from two Greek words meaning "scale wing." These insects with scales on their wings belong to the *Lepidoptera*. The milkweed butterfly has these scales: so it is a member of this order.

So are all butterflies. So, also, are all moths. The butterflies and the moths make up the principal divisions of the order.   Butterflies for the most part are the scale-wings that go out in the daytime, and the moths are usually the scale-wings that go out at night.   There is another fundamental difference, the scientists will tell you, between these cousins.   The wings of butterflies are put on with hinges like the shutters at your window.   They can be brought together over the back or they can be left sticking out at the sides, but they cannot be folded up.   The moth, on the other hand, folds its wings together and lays them down on its back.

And of the butterflies the greatest of them all are those colorful, dull red and black adventurers, four or five inches across, that live by the milkweed, that are sometimes called monarch butterflies.   It is this particular scale wing that we are talking about.

I do not suppose that there is anybody who has not at some time caught a moth or a butterfly in his hands. At those times he always finds a sort of dust remaining on his hands.   This dust is made of wing scales.   These scales are placed on the wings a good deal as the scales of a fish are put on, or as slate roofing is laid.   They come in different colors, and Nature often works out fancy designs on the wings of the butterfly by arranging them very carefully, just as the builder works out fancy designs on the bath-room floor by fitting in tiles of different colors.

But here is another odd thing that scientists have found out about these scales.   They are made of the same material as is hair.   A scale is merely a hair in another form.   The body of the butterfly may have both hair and the wing scales on it.   It is only two ways of using the same building material just as a disc or a spike might be made of steel.

But to get back to the milkweed butterfly. The first remarkable thing about it is the fact that it is a stupendous traveler. It migrates from south to north and back again from north to south, all in a single season. A single milkweed butterfly may unfurl its gaudy wings in the vicinity of Mobile, Alabama, early in the spring, and in August may be flitting idly about on the shores of Hudson Bay. The cold may begin to nip and it will turn south, and November will find it again on the Gulf Coast.

CATERPILLAR OF A MILKWEED BUTTERFLY.

There are many migratory birds, but few migratory insects. This single family of butterflies, however, probably furnishes more living creatures that travel back and forth across wide areas than do all other animal families combined.

When the tender shoots of the milkweed first appear in the far South, very early in the spring, they are soon visited by certain mother butterflies who, the previous autumn, had come down from the North that they might live comfortably through the otherwise unhappy winter. These butterflies lay their eggs on the under side of the leaves of the milkweed.

This butterfly here gives its first evidence of a sense of color and design, a taste in which matter follows it through its entire life cycle. This dainty egg is a glistening, amber green. It hatches into a caterpillar, to which is given a stupendous appetite like all of its kind. It soon becomes a sleek and ample specimen two inches long, radiant in brilliant bands of yellow and black—a most conspicuous creature. The chrysalis into which the caterpillar hardens for its sleep before becoming a butterfly is, again, of emerald green with gold and black dots, hung by a silken black thread in a bag to hatch. Then the butterfly itself appears with its orange-red wings, tipped with black and dotted with white.

This taste for brilliant colors, it will be found, plays an important part in this creature's game of keeping itself alive in a worldful of enemies. Follow its career and see how it plays a vital part in its life.

But here I must pause to call attention to this marvelous transformation which takes place while this creature is in the pupa or chrysalis stage of its development. The butterfly, like most insects, finds the egg to be a convenient form in which to launch its offspring into the world. It lays the egg in the midst of much food, so that the caterpillar may begin eating as soon as it is hatched.

The chief business of the caterpillar is to eat and grow. If a man had an appetite as large in proportion to his size as that of a caterpillar, he would devour a hundred pounds of food at a sitting. This caterpillar is but a fat, sluggish worm, full of plant juices, having no purpose of its own. Its mission is to store up within itself the materials from which a butterfly may be made.

When it is well stocked with these materials, it enters the hardened state, becomes a chrysalis, and hangs itself up to ripen.

It is during this period of quietude, of sleep, that the miracle is performed. The materials in the caterpillar, this sack of milkweed juice, gradually begin to take form —to become a butterfly. Legs and antennæ begin to shape themselves. A well proportioned body of three sections evolves. Four gorgeous wings, still much compressed, are spun against the time of coming forth. Those delicate hues that are to spangle them are extracted from this store of juice. The shingles that are to make them, that are laid as carefully as the mosaics of a cathedral, are created as though selected by the eye of a master artist. This caterpillar, it seems, knew just what materials were necessary for this structure. Eventually everything within the chrysalis is used up. There are no odds and ends left over. Then the butterfly emerges, radiant and dazzling creature of the sun that it is, a masterpiece of nature.

The eggs laid on the early milkweed of the South pass through their development stages and become young butterflies, and these flutter away in a seemingly vagrant manner, but actually join their mothers in a drift to the North. Spring goes north slowly, carrying the butterfly and calling forth the milkweed as it goes. Wherever a milkweed flings forth its leaves, there is likely to have arrived that very morning a butterfly from the South that will lay her eggs upon them. By doing so she adds to the numbers of the army that is steadily drifting northward.

It takes spring about two months thus to span the United States with young milkweed. It takes the milkweed butterfly two months to cross from the Gulf to Canada. A man might have walked the distance as quickly.

This silent army has made its pilgrimage almost without being noticed. Even the careful observer can hardly see that butterflies travel to the northward—they seem to fritter about so aimlessly. But the result is convincing. There are many butterflies to be found even beyond the point where milkweed grows. Some of them seem not of the season in which they are found, their coloring suggesting that they are more than one year old. Have these same individuals, one may ask, made this round trip before?

So numerous have the hordes grown, however, that, when the drift south begins, the fact is more obvious. In Pennsylvania, in September, for instance, they have appeared in swarms that have taken days to pass. They have flown for hours over St. Joseph, Missouri, in the same month, three or four hundred feet up and so dense as to cloud the sky. A month later they appear in Florida and in Mississippi. They have reached home from their summer excursion.

That it might make this round trip, the milkweed butterfly must be equipped in various ways. Primarily it must be able to supply itself with food. For this purpose it is provided with a very unusual instrument. This instrument in ordinary operation would appear to be none other than that straw with which human beings are prone to take their soft drinks at soda fountains. The butterfly drinks through a straw.

This straw may be as long as the body of the butterfly itself. The mystery is where the creature has been carrying such an instrument without your ever having noticed it before. But if one watches carefully he will solve the mystery. The insect may select the blossoms of the clover, thistle, or goldenrod when lunch time approaches. It is not a particularly well mannered diner and quite loses its poise as the thought of food possesses it. It is then that its drinking tube begins to appear. Under ordinary circumstances this is wound up like a watchspring and cannot be seen. But at thought of food it is excitedly unwound and rewound. It disappears and then is extended to a surprising length. The insect beats its wings in high excitement. Then it reaches into the honey of the flower and partakes of deep draughts. The wings work more and more slowly; then come to complete rest. It is satisfied.

But to get back to that graver situation, even than that of foodgetting, fraught with a bigger chance of tragedy, which the butterfly, in common with all insects, must meet. The butterfly throughout its life cycle is a creature of peace, a feeder only upon the flowers. But there are all around it hundreds, thousands, of other kinds of creatures which live through eating, not vegetable life, but other animal life.

So, when the mother butterfly first lays her eggs, she knows that spiders and crickets are fond of them and that a considerable part of them will be devoured. She lays more than enough to discount this prospective loss.

She knows that, when her young has grown to the caterpillar stage, is fat and helpless, there will be many creatures ready to nourish themselves on its well fed body. Most deadly of these are certain flies, called ichneumon flies, which do what seems to be a most fiendish thing. They insert their eggs in the body of the helpless caterpillar. There these eggs hatch into grubs and these begin to eat and grow, eat and grow, until the poor caterpillar dies from having its vitals devoured, of which event its unwelcome guests take no notice, but continue their feasting on its body. So are the careers of many butterflies nipped.

Then there is that other danger which is great to the caterpillar and which follows even the grown butterfly all the way to its final end—the menace of the bird life that is everywhere so abundant and much of which depends on catching insects as a means of livelihood.

The chief device of insect life in protecting itself from the birds that prey is through so imitating its surroundings that it cannot be seen. Most varieties of butterflies use this method. Their caterpillars look so like the plants on which they live that it is hard for the birds to see them. The butterflies themselves so fade into their surroundings that they do not attract the attention of the birds that might otherwise devour them. Those types that do not fail to escape detection are devoured and cease to exist. Those types that do succeed in outwitting the birds live and multiply. So does cleverness have its reward.

But now appears the milkweed butterfly and seems to give the lie to this whole theory. Its coloring and that of its caterpillar is the most gorgeous of them all. They do not imitate the plants on which they live at all. The caterpiller is the biggest, fattest, apparently the choicest prize of them all from the standpoint of a hungry bird. It makes no attempt to conceal itself. On the contrary, it attires itself most gorgeously and goes to great pains to advertise its presence.

Yet it is not devoured. It is the most abundant and widespread of all the butterflies.

Back of this situation there is one of the most alluring sets of circumstances in all insect life. It is based upon the fact that this milkweed butterfly, both the caterpillar and the adult, this monarch of its race, this venturesome traveler, has a peculiar quality not possessed by its fellows. It is a hateful and repellant quality. Its body secretes a pungent fluid that has a rank odor that tastes bad to birds, nauseates them, makes them sick.

The ordinary varieties of butterflies are sweet and tasty to the birds, but this gaudy peacock of the race is most repulsive. It is because of the possession of this unattractive quality that the monarch survives.

The dominance of the milkweed butterfly probably came about in the following manner: Butterflies at first were all about the same. They and their caterpillars looked a good deal alike, all mimicked their surroundings to keep the birds from getting them.

Then they spread out and began to develop peculiarities. One sort of butterfly came to taste slightly different from another. The birds did not like it as well. They began to avoid it. This they could not always do because it did not yet have any very outstanding marks of difference from the other butterflies.

There were some of these bad-tasting butterflies, however, that looked different—had marks of color peculiar to them. The birds could recognize these and avoid them. Those that still looked like ordinary butterflies, however, were still killed by mistake.

The caterpillars and butterflies with colors that advertised their bad qualities developed more rapidly.

THE MILKWEED, OR MONARCH, BUTTERFLY.

The more brilliant their colors, the surer could the birds be of avoiding them. The more conspicuous they were, the safer they were. Those colorings were advertisements of their inner unattractiveness.

So, as the centuries passed, was a race of butterflies developed that survived because it became all the time more and more conspicuous. The large and meaty members of this race were left alone by the birds; while the large and meaty members of other butterfly tribes were sought out and devoured.

So did this race grow lustier than its fellows. It might with safety fare afield. It might go on its travels, for the creatures that preyed on butterflies in general would not molest it. It might frolic in the sunshine of summer from the tropics to the far North. The modest and sweet-flavored butterflies, however, must hunt the shadows, must assume the colors of the plants among which they lived, must sit still and prim, for motion attracts the eye of the huntsman bird.

So did the butterfly with the bad taste flourish beyond all others. It flourished chiefly because it advertised its inner unattractiveness.

Then another odd thing happened that was based on the safety from attack of the milkweed butterfly and the gaudy colors that it wore to advertise itself. The coloring of other caterpillars and butterflies might happen to be a little like that of the milkweed variety. As a result the birds would tend to be suspicious of them, and would avoid them. Thus would the imitating varieties survive, while those that were frankly good to eat would be destroyed.

So did it come to pass that other races of butterflies were developed that were colored like the milkweed butterflies, but which were without their bad taste. To be sure, they were rather unimportant races that hung on the fringe of the monarch of the milkweed. But they were resorting to a strange device for their protection—a device that was based upon another strange device. They were successfully playing this desperate game of life and death in a way quite their own and almost without parallel in all the sweep of Nature.

This milkweed butterfly, monarch of its race, seems to be an idle, truant fellow, wandering about the world and serving no purpose whatever. It would probably be a great mistake, however, to condemn it hastily. Milkweed, for instance, might entirely overrun the country, the United States might become a jungle of milkweed in ten years if the butterfly caterpillars did not keep it down by eating it.

Whenever milkweed begins to get more than usually plentiful, it furnishes food for more butterflies than usual, and they increase in numbers and their caterpillars devour it. On the other hand, when the milkweed gets scarce, there is shortage of butterfly food, and they, in turn, become scarce. This gives the milkweed a new chance to develop.

Thus does Nature tend to maintain a balance as between butterflies and milkweed. Each helps the other when it tends to become too scarce. Each restrains the other when it tends to become too abundant. So does Nature manage to keep on hand about all the stock of milkweed and of butterflies that it deems wise. Without the butterflies this milkweed would run riot. Without the milkweed the world would have no monarch butterflies. As it works out it keeps about as many of each as it ought to have.

QUESTIONS

1. (a) Nearly everybody has seen birds migrating, but few people know that some insects obey this instinct. Why do we regard the milkweed butterfly as the traveler of the insect world?

   (b) Describe its journey—the time of the year it sets forth, how it supplies itself with food, the distance it covers, and the goals it reaches.

2. How can you tell a moth from a butterfly? If you were making a collection, could you catch both members of the Lepidoptera at the same time of day?

3. (a) What form does the butterfly take when it hatches out of the egg? Is there anything unusual in the manner of its development?

   (b) Throughout its life history, the monarch displays a find taste for color. Why does this insect flaunt its gay stripes to the world?

4. (a) How does mother butterfly keep spiders, crickets, and ichneumon flies from killing her offspring?

   (b) Many insects are hard to see because of their colors and so escape their enemies. Is the monarch more inventive? How does it meet the bird problem from the very start?

5. In thinking of this daring butterfly, its size, travels, food habits, its methods of avoiding its enemies, are there reasons for considering the "monarch" well named?

6. (a) In what way does the monarch help the farmer?

   (b) How does Mother Nature maintain a balance between butterflies and milkweed?

CHAPTER V

THE MAY FLY

THE nearest thing to a fairy in all the animal kingdom is the dainty, dancing May fly, which comes into the world for a day at the approach of summer, frolics itself to death in the sunshine, and is gone.

One moment an ugly creature of the mud, the next, as though by the gesture of a magic wand, it is off, radiant as an oriole, with a shimmer of new-found wings, fragile as the gossamer of a spider's web, with its streaming three thread-like tails, twice as long as its body, marking time to its fitful, up-and-down, joyous flight.

Out there in the blue, as by instinct, it finds its mate, for this is its bridal day. The two frolic together for an hour in a world that seems too gross for creatures as fine as they. Then they part, and each, ceaselessly, goes on and on until the last bit of vitality in its frail body is spent.

Then it settles, perhaps on your coat sleeve or mine. The feet with which it clings are unused little feet, for all its life it has been in the air. They are hardly able to support even its dainty weight. It wavers there for a moment and tumbles over like a bit of thistledown. Its brief span of life has come to an end. The May fly is dead.

In the books of science the May fly is called Epheme-rida, which means "the child of a day." It lives but such a span and is gone. In fact, there are many of its kind which do not remain so long. They come out with the declining sun of a May afternoon, dance through the mellow evening, exhaust themselves by the coming of midnight, and go the way of the fathers.

So brief is its life that one who would make its ac-quaintance must hurry. The opportunity comes but once a year—in the spring—and is not unlikely to im-press itself on the individual who lives near a lake or a river, for the May fly is water born, and near borders of it it is likely to be very plentiful. For one day or two or three in the year the lake dweller is likely to have the existence of the May fly impressed upon him. The street light in his town may draw millions of them from the darkness roundabout—millions that happen at that moment to be engaged in the fling that comes in that short time when they are given wings. The street lights may be darkened with them, the pavements made slippery with their crushed bodies. The street lights of Atlantic City gather, each season, a countless toll of May flies. So do those of Niagara, and Natchez on the Mississippi, and Geneva on its far away Swiss lake, and ancient Bagdad, where the Tigris flows down to the sea. Along the St. Lawrence and on the shores of the Great Lakes, May flies swarm by the million so as to produce imitation snowstorms in the summer. The air is thick with them. They become a nuisance, but a nuisance so short-lived as not to be a great hardship.

A strange creature is this dainty May fly. One of the

strangest things about it, again quite fairy-like, is the fact that it eats nothing from its cradle to the moment of its death. Thrust into a world where nearly every living thing is constantly in search of food, the May fly goes its way with never a thought of food. Surrounded by creatures that lie in wait for each other and devour each other, the May fly eats not nor thinks of eating. It does not even possess a mouth with which it might eat if it felt so disposed. It comes into the world with a certain amount of vitality in its frail body, it dances gaily until that strength is gone, and dies.

Equally strange is the birth of the May fly, a birth which almost any observer may witness if he happens to be sharp-eyed by the waterside almost anywhere at just the right time. If he watches closely, he may see many tiny creatures as big as house flies emerging from the mud at the bottom of the lake or stream. These little creatures are in the act of doing a very unaccustomed thing. They are leaving their usual place of abode, where they have stayed close at home for two or three years, and coming to the surface—a thing they have never done before. Perhaps they merely float to the surface of the water, or perhaps they crawl up on some stone or twig at the water's edge.

These are ugly little water insects with sturdy legs, an active tail used in swimming and with gills through which they breathe as do other water creatures. But as they reach the surface of the water, a most surprising thing happens. Their skin down the back splits open, there is a wriggling within it, and, quickly as a flash, there emerges, not this water creature in a new dress, but the dainty

May fly, a thing as different from the water insect from which it came as a peacock is different from a rat. In a way it is more different, for but a moment ago this creature breathed water through gills like a fish, and now it breathes air and would drown if you thrust it under water.

This newborn creature stretches its gossamer wings and flutters away to a nearby twig. There, in a very few minutes, it again sheds its skin, even to the covering on its wings. It is strange in this also, the scientists say, for it is the only insect under the sun which sheds its skin after it has acquired wings. Then it flies away on its endless dance until the time of death.

Back of this brief blaze of glory in which the life of the May fly goes out there is, however, a long period of preparation. The little water insect that gave birth to the May fly has been battling for its existence for one or two or possibly three years among the creatures that live in the mud at the bottom of the water. The original egg from which this insect came was laid by a May fly two or three seasons ago. It sank into the mud and there hatched into, first, the larvæ, a crawling and wriggling thing, which kept growing and developing into higher forms.

These insects have difficulties peculiar to them as they grow. Their skins or shells will not stretch, so they grow inside those skins until they become too tight. Then they burst open, and a new and looser skin forms, which is worn until it again becomes too tight, when it likewise is burst and discarded. As skins are discarded there are often quite pronounced changes in the nature of the insect. The larvæ, for instance, later takes the form of what is known as the nymph, and this nymph emerges into the full grown insect. These three forms are common to many insects.

The nymph of the May fly is a very sturdy creature. It lives in the mud and eats, chiefly, decayed vegetation, but is not scornful of an opportunity to devour the other insects round about that are smaller and more helpless than itself. It, in turn, has many enemies. Chief of these is the nymph of the dragon fly. The dragon fly, or mosquito hawk, or snake doctor, is, as you know, a very forceful character much given to battle. Despite this, it is in reality a cousin to the delicate May fly. Its nymph lives side by side with the nymph of the May fly in the bottom of the stream. Even in its youth the dragon fly is already a dragon. Its nymph lies there in the mud, all quiet and harmless looking. If a lesser swimmer, like the nymph of the May fly, comes too near, however, a great transformation takes place. What appeared to be an innocent shield over the front of its head unwinds and extends itself like the telephone rack in a business office. It reaches out an inch or more. On the end of it is a nipper. This nipper lays hold of the lesser creature, the nymph of the May fly, for instance. Then the extension arm again folds itself up, forming again the shield, but the May fly nymph is back of it and is even now being greedily munched by the stout jaws of the dragon.

The dragon nymph eventually comes to the surface, splits itself open, and turns into a dragon fly. That dragon fly, however, is a far different creature from the May fly. It goes very energetically about its business.

It eats heartily. It is a great huntsman. It gathers its legs into a sort of net and darts swiftly about. It scoops such insects as gnats and mosquitoes into its net and devours them. It lives long and bags much game.

It is on this day of frolic that the May fly lays its eggs. Through all the three years of its cycle this is the only day upon which it takes thought of its own reproduction. It is undoubtedly true that Nature transforms the mud-inhabiting grub and turns it into a creature of the air for this brief period that it may be given an opportunity to scatter its eggs far and wide.

While on the wing the May flies spread out over the surface of the water. Those of them that strike inland, that congregate about street lamps, fail of their purpose, have lived to no avail. As their more successful fellows float so waveringly up and down in the sun over the water they are directed by an instinct to plant their eggs where they will have a chance to hatch out into other generations of May flies. Some of them drop their eggs and flit about a while longer until their strength is gone and drop to watery graves. Others fold their wings about them and plunge into the water, diving deep to assure their eggs the right opportunity, and in doing so sink to their deaths. Others, myriads of them, fall on the surface of the water and there perish, but not before they have laid their eggs. It is the way they want to die— in the performance of their purpose. Their final mission is accomplished, their day of glory at an end, their dance of death completed. What could be more fitting than that they rest upon the water that so long sheltered them and is to shelter their children and children's children.

Along the shores of Lake Ontario on a May day, the bridal May day of the May fly, there are windrows of these fairy forms, brought in by the waves like windrows of hay miles in length, all having died as they should while planting their eggs at the end of that one wild day of joy that caps an otherwise dull, drab life.

Watching the life of this frail and delicate little creature run its course one would be likely to conclude that here at least was one thing beneath the sun which served no useful purpose. The May fly, it would seem, is too fairy-like, too frivolous, too weak, too unimportant to serve a material end in this work-a-day world.

THE MAY FLY.

But this is not at all true. It has one rather important field of usefulness, a field in which it adds a good deal to the pleasure and to the food supply of that lordly individual who calls himself man, and who is likely to think that the purpose of the universe is to supply his needs.

The May fly, its nymph, and its larva are the chief items on the bills-of-fare of many of the most important food fishes. Because of the purpose it serves in this connection, it is sometimes called the shad fly. It is to the food fish families of many fresh waters what rice is to the Chinese or hay is to the horse. They eat it all the time. They eat the larvæ when they are little, the nymphs as they grow bigger, and finally feast to bursting on the May flies themselves when, after their final dance, they fall in such quantities on the surface of the water.

It is hard to find out whether there would be fishes in the streams and lakes without May flies. Probably they would find other food in its place. It may be, however, that they would not, that anglers would whip the streams in vain when they go out with hook and line if this tiny insect had not kept a dependable supply of food on hand for the fishes through the cycling seasons. Fishermen and lovers of fish as food may owe much to the May fly.

QUESTIONS

1. (a) What is the author's attitude toward the May fly? Do you agree with him in calling the May fly "the fairy of the animal kingdom"?
   (b) Is this insect's name well chosen? What other titles might you give the May fly? Give reasons for your choice.
   (c) Read lines which bring out the "fairy-like" nature of this insect.
   (d) Do you think you could compose a little poem "on these creatures of a day"?
   (e) Draw a picture of these "fairy ships sailing in the sea of air." Locate your scene in a place suggested in the story. Keep in mind the old Dutch saying, "As thick as May flies."
2. Tell in your own words the part of the story which interested you most.
3. Explain with illustrations these expressions:
   (a) "Radiant as an oriole."
   (b) "Fragile as the gossamer of a spider's web."
   (c) "A world too gross for creatures as fine as they."
   (d) "Is not scornful of an opportunity to devour."
   (e) "In the performance of their purpose."
   (f) "Like a bit of thistledown."

    (g) "Brief span of life."

    (h) "Go the way of the fathers."

    (i) "Brief blaze of glory."

    (j) "That caps an otherwise dull, drab life."

4. The story abounds in beautiful imagery. Read aloud the passages that please you most.

5. What do you like best about this selection—the narrative, the style, or sympathetic treatment of the subject?

6. Mental moving pictures. Plan a series of mental moving pictures, for example:

    (a) A day under water with the grub.

    (b) The approaching enemy.

    (c) The sturdy nymph battling for its life.

    (d) Voraciousness of its cousin, the dragon nymph.

    (e) Coming to the surface, losing its gills, and becoming an insect of the air.

    (f) The May fly shedding its skin again on a nearby twig.

    (g) The brief, blissful bridal day.

    (h) Fulfilling its life's purpose.

7. Think of the insects we have studied so far. Arrange your report in this way:

Name of Insect: Order	Characteristics	Helpful to Man	Hurtful

Volunteer Work

8. Question for informal debate. Speeches to be limited to two minutes each.

    (a) Which makes the greater contribution to man's welfare, the fig insect or the May fly?

    (b) Resolved, "That the May fly's day was not lived in vain."

CHAPTER VI

THE LADYBIRD BEETLE

ICERYA was the little immigrant insect that upset the balance of Nature, which, on the face of it, would seem to have been a good deal of an accomplishment.

That name, Icerya, is foreign sounding, but, at that, one would expect, if he could find out from whence it came, that it would give an indication of the sort of person its owner was likely to be. Names, in their beginning, usually have a way of being descriptive. The original Mr. Strong, of course, had a great deal of muscle, and the original Mr. Sheppard looked after the flocks. But Icerya, it seems, came by her name because a certain Dr. Icery, little dreaming of the history she would some time make, happened to discover her.

The first Icerya to reach American shores was not even a legitimate immigrant. She was a stowaway. She came in by the Golden Gate at San Francisco. Being only a tiny insect, she was hidden in the branches of an orange tree that was being brought all the way from Australia that it might be planted to grow in the balmy climate of the west coast.

The orange tree was set out and prospered, and Icerya, the insect which is the scale, likewise grew and multiplied. The marvel of it was the rapidity with which it increased.

The little Icerya are mite-like creatures, barely big

enough to be seen, that run actively all over the tree that is their home. They do not have to worry about food, for, to satisfy their hunger, they have but to stop on any leaf or twig, drill a hole with an instrument they carry for the purpose, and drink their fill of sap.

Like all growing insects they wear their skins until they get too tight, whereupon they split open down the back, drop off, and the creatures then proceed to grow and fill out the new coats they acquire. When coming out of their old clothes, their gummy bodies are exposed to the air and harden into a crust.

The male and female mites look alike when they are little, but as they grow older they take different courses in their development. The male finds that he has acquired new abilities with each moulting. Eventually he is a very active creature all decked out in wings upon which he can visit all about the orchard. By the time he is ready to start a family he is probably in some tree far away from that in which he grew up, and like as not he marries some woman scale whom he hardly knows.

The female, in shedding her old coat, finds that a very different thing has happened to her. In dropping her skin, she finds she has likewise lost her legs. She can no longer run about. Since she is to stay all the time in the same place, she no longer needs eyes, so these also are discarded. That leaves her, a plump little body sitting on a limb with nothing much to do. She still has her drill, so she sinks it into the twig she is on and occupies herself with drinking sap.

Then, also, she must raise her brood of young ones. It is these mother insects grouped together that become the scale—cottony-cushion scale it is called because of its appearance.   That this little mother may build a house for her family she secretes from her body the material for making it.   This is a sort of white wax.   It is the outside appearance of this house that causes her colonies to be called the cottony-cushion scale.   Inside this house she lays her eggs and dies.   The mites hatch out, and another generation is begun.

Now just this sort of thing had been going on for a million years in Australia, and no harm came of it.   But it got started in California because of this one Icerya stowaway, and in a few years there was a great hulla-baloo.   The cottony-cushion scale had spread to all the orange groves in the southern part of the State.   On every tree in all the groves there were so many mother Iceryas, each with her drill running down into the sap streams, that they were drinking them dry.   After they got their sap there was none left to nourish the tree, to grow its leaves and fruit.   As a result all the orange trees of California were threatened with death.

The scientists surveyed the situation and came to the conclusion that the balance of nature had been upset. This scale in Australia, they concluded, because they knew the ways of insects, had an enemy of some sort that kept it within bounds.   It had been brought to the United States, but its enemy had been left behind.   It was as though milkweed had been taken to Australia, but the butterfly which feeds upon it and keeps it from becoming too abundant had not gone along.

This cottony-cushion scale was running away with the situation.   It had broken up the balance of Nature because it had left its natural enemy away over on the other side of the Pacific Ocean.

The government is always as ready to hurry to the assistance of its citizens in distress of this sort as the fire company is to put out an unexpected blaze in their houses. It works upon the theory that it will help the citizen on tasks that he cannot perform for himself. The individual citizen cannot put out a burning house, nor can he solve these riddles of science. These are, therefore, matters with which the community or the government should help. So in the case of the insect invasion the government dispatched a wise man to Australia, told him to find the natural enemy to this scale which was killing the orange groves, and hurry it to California.

In Australia the search did not take very long. Among the orange trees, it was found, there was to be seen, here and there, a bit of this cottony-cushion scale. It was here that Icerya made her home. There was not enough of it, however, to hurt the trees. Nobody paid any attention to it. Nature had established a balance. This American went into the orchards, watched carefully about the places where the scale was to be found. One day he witnessed a strange proceeding.

A monster came into one of the scale colonies. This monster, from the man standpoint, was about as big as half a holly berry, round-backed like a turtle, and hard-shelled. It was rather dashingly colored in black and red. From the standpoint of Icerya it was a towering, bone-crushing dragon, slathering its greedy jaws with the torn bodies of its victims.

The scientists saw this intruder crash through the wax-built houses, as a hippopotamus might small chicken coops, root around among the ruins, pounce upon the form of every little scale insect it saw, and devour it. A creature of vast appetite was this marauder. It went on and on from one clump of white tents to another, eating its fill. It seemed a demon of a single purpose. There was only one thing it sought. It hungered for a single food—scale insects.

This scientist knew he had found the object for which he had crossed the Pacific. Here was the creature that held Icerya, the cottony cushion scale, in check. Here was its natural enemy that had, through the centuries, kept it from becoming overabundant in Australia. Here was the balancing check that it lacked in the United States. Over there it had escaped its natural repression. It had enjoyed an unnatural freedom. It had developed more rapidly than it should. Nature was out of balance. There was a riot of scale insects.

To be sure, Nature would find a way to restore a balance even if this enemy were not taken to America. So abundant were the scale insects that, in a few years, they would kill all the orange trees and then they would die for lack of food. This would be bad, even for them. It would be very expensive to an outside, interested party, man. Man was taking thought to avoid the necessity of so costly a method of bringing Nature back to a balance. He was using his knowledge of these small insects to his own good.

The odd thing about it was that this creature, peculiar to Australia, did not seem at all strange to this American scientist. In fact, it would look like an old friend to almost any American boy. It appeared to be none other than that person to whom the nursery rhymes addressed that couplet:

"Lady-bug, Lady-bug, fly away home,
Your house is on fire and your children will burn."

But in Australia they called it "ladybird," the lady-bird beetle. It is a strange name for an insect, tying back, it is said, to Our Lady, the Virgin Mary. Old records show that ladybirds were creatures held in high and somewhat religious esteem in ancient Scandinavia and England and, in fact, throughout Europe. Men did not then know how well they were served by these small creatures, but everywhere there seemed to exist an instinctive affection for them.

The scientists called this Australian ladybird Vedalia, which, again, is a bit of a fancy name with no particular significance. It was like the lady-bug of the United States in general appearance, but had certain peculiarities. Its dress was a polkadot. The ground of this was often of red with black spots, but sometimes of black with red spots. Different families of ladybirds had different numbers of spots on their backs.

The ladybird is a beetle, which means that she is a sheath-winged creature, a coleoptera. The beetles are different from all other insects, with a single unimportant exception, in the fact that they have these hard, shell-like wings. The humble tumble bug, for instance, is a beetle, as are the snapping bugs that have got many a country schoolboy into trouble, and the apple twig borer of the orchard.

Most beetles eat vegetable matter and are destructive of crops, food, or wood. They are mostly a nuisance from the standpoint of man and are harmful to him. The beetles of this lady-bug family, however, even those native to the United States, have an odd taste in food. They eat but one sort of food, which is scale and similar insects. There are many sorts of scale that grow on trees, rosebushes, broom, even on grass, and the lady-bugs and and the ladybirds devote themselves exclusively to eating them up.

Thus they are of benefit to man. It is doubtful if he would be able to grow fruits without them. They are the beetles toward which he should feel most friendly. He was acting in his own interest when he informed the absent mother of the danger to her household from fire.

But this little ladybird had lived a long time in far away Australia, entirely cut off from others of her kind. She, as a result of her surroundings, had developed food habits which, it seems, are different from those of any other creature. She fed exclusively upon cottony-cushion scale. She had been doing it for a million years. She had no other idea of food. When she could not get cottony scale she went without food, even to the point of starving to death.

Yet when this cottony scale, Icerya, came to America and grew in such abundance that it was destroying the groves, the American cousin of this ladybird, a scale eater also, and so like her that they could hardly be told apart, ignored it as a source of food. It might suck all the juice from all the orange trees in California for all the American lady-bugs cared.

So this scientist packed up little boxes of these Vedalia ladybirds with polkadots on their backs and shipped them hurriedly to California where Nature was out of balance. One immigrant was sent in pursuit of the other. Vedalia should discipline Icerya. No sooner had the pursuers arrived than they were released in the orchards where the Icerya were thickest. They were hungry after their long journey. Their appetites when normal are such as would seem impossible to creatures so small, but after the long fast it seemed as though they could not get enough.

After feeding, each of these females went forth and laid 300 eggs which, under these favorable circumstances, practically all hatched. The little grubs that resulted immediately began devouring scale, all grew to maturity in the presence of this abundant food supply, and the 150 females among them each laid 300 more eggs. Under these favorable circumstances, the scientist figured with his pencil that a single imported mother would have 75 billion offspring in five months.

The results of the experiment were nothing less than marvelous. In six months the ladybird bettles had increased in such numbers that they had spread all over Southern California, had turned the tide against the enemy, had so reduced their numbers that the trees again burst into leaf and bloom and proceeded to produce crops of oranges. In a year the scourge was suppressed. The balance of Nature had been restored.

This drama of insect life has since been worked out at a number of places in the world. The cottony-cushion scale, originating in Australia, has, at different times, found its way to New Zealand, South Africa, Portugal, the Hawaiian Islands, Italy, Syria, Egypt, and France. There it has attacked orchards just as it did in the United States. The United States, after its successful experience, has become a handier source of information and of supply for ladybirds than Australia, and most of these countries have come here when in trouble and we have furnished the cure. The State of California maintains at Sacramento a barracks of these ladybirds. They are kept in condition, all the regiments recruited to full strength, ready for immediate service. When an outbreak is reported at any point, even though it be as far away as the shores of the Mediterranean, or the South Pacific, their bugles are sounded and they are off to repress it just as the regular army used to put down outbreaks of Indians here in the West.

This case of the scale and the ladybird is regarded as the perfect demonstration of the manner in which the balance of Nature can be upset by some act of man like that of bringing the Australian scale to America without its controlling enemy, and of the steps that may be later taken by man, as a result of his understanding of insect life, toward restoring that balance.

The world undoubtedly owes a great debt to the ladybird beetle, for she and her cousins have saved many fruits and shrubs besides the oranges from scale destruction. She stands forth as man's best friend among her kind.

Beetles are generally enemies of man—tend constantly to injure him. There is one ladybird beetle, for instance, with habits quite different from those of Vedalia. Instead of living on scale insects, she lives on beans. The government has found it necessary to campaign against her to save the bean crop. There is the saw-toothed grain beetle, the carpet beetle, the weevil, the larder-beetle which eats cheese and such, the woodborers which destroy vast quantities of timber, the corn-borers, June-bugs which injure the lawn. There is the rhinocerous-beetle, giant of its race, which grows two and a half inches long in this country and six inches long in the West Indies and digs among the roots of sugar cane and corn to their injury. There is the potato bug which all farmers must fight, the elm-borer, the maple-borer, and hundreds of others.

All these are enemies of man. In fact, it must be constantly born in mind that most insects are his ene-mies—that there is constant warfare between man and the insects. One has but to recall the battle with the boll weevil in the South, with grasshoppers in the West, with mosquitoes which carry yellow fever, with house-flies that scatter typhoid, to realize how desperate is this fight. But through it all the scale-eating ladybird is his constant and vigilant helper, ever ready to put her most unusual appetite, and her most unusual capacity for re-cruiting armies of her kind, at his service. She is his friend. With her help the United States Department of Agriculture may control the spread of the San José Scale, so called because it was in the large orange-growing center of San José, California, that it began its destruc-tion. With her help, the damage done by the cottony scale, one of the 37 insect pests introduced into the United States from foreign countries which left their natural enemies at home, is rapidly growing less.

QUESTIONS

1. (a) How did the Icerya, mite-like creatures, find their way into California?
   (b) Were they welcome guests?
   (c) How did they make themselves at home in the orange groves of California?   On what did they feed?

2. (a) How does Mother Icerya spend her time after she loses her legs and eyes?
   (b) Describe the cottony-cushion scale.

3. (a) In what way did the cottony-cushion scale in California disturb the balance of nature?
   (b) Did the American scientists sent by the United States to Australia, home of the scale, find it doing much harm?   Why not?

4. (a) Are beetles friends or enemies to man?   Tell about the Australian ladybird's discriminating taste in the matter of food.
   (b) America had lady-bugs on its own soil.   Why was it necessary then to import the cousin of this insect from Australia?

5. (a) Describe the enemy of the cottony-cushion scale.
   (b) How long did it take the ladybird beetle to conquer the scale.
   (c) Were you surprised to read of the grandchildren of a single imported Vedalia mother in five months?
   (d) Why should our boys and girls be glad that Vedalia was able to discipline Icerya?

6. (a) In days of yore our regular army was often called upon to put down outbreaks of Indians in the West.   How do ladybirds kept in the barracks at Sacramento by the State of California police the plant world and repress inroads of the cottony-cushion scale?
   (b) Resolved: That all plants coming into the United States should be as carefully inspected as are immigrants applying for admission to the United States.

Chapter VII

THE GIPSY MOTH

 MOST cruel and frightful war has been raging right here in the United States for nearly fifty years and is likely to go on for as many years more.

It started in the midst of Massachusetts, not far from Lexington and Bunker Hill, where the Revolution had its beginnings. The onslaught was successful, and the invaders took possession of two-thirds of New England. After fifty years of fighting they had been driven back until the battle line, like that which the Germans once maintained from the Alps to the sea, ran irregularly from Portland, Maine, to Providence, Rhode Island.

The invader was an insect, the gipsy moth, which came from Europe. Its determination was to eat all the leaves off all the trees in New England, particularly those of the oak and birch. Since leaves are the lungs of trees and they cannot live without them, death threatened them all. Having conquered New England, the invader might be expected to push on to the west. How far it would go would depend on the opposition it met, on the skill and tactics of the defenders. The issue is not yet decided.

This is one of the greatest insect wars that the world has ever known. There have been other such wars, as,

73

for instance, that in which the ladybird put the orange tree scale to rout. The boll weevil in the South has fought desperately for the destruction of the cotton crop, but for a shorter time. There have been outbreaks of grasshoppers, but they have soon passed. Never before has man set himself to battle with an insect, applying all the scientific knowledge that the world has to offer, bringing armies of other insects from many parts of the world, marshalling his forces, mapping his campaigns, carrying on decade after decade.

Like many another war this one had its beginning in a very small incident. Away back in 1869 a scientist of Medford, Massachusetts, was conducting experiments with moths. The silk worm moth is important to man since it furnishes him with the material for making certain dainty articles of clothing. This moth was suffering from plague. It had some disease that threatened to kill all its kind. This scientist was attempting to cross the silk worm moth with some other sort of moth and get a new creature that would spin silk and still be proof against the plague.

So he brought to this country certain hardy moths of Europe. Among them was the gipsy moth, well known in England, throughout Europe, and, in fact, even in far-away Japan. It was called the gipsy moth because the male was yellowish brown, about the color of the face of a gipsy. In France, oddly, it was called the zigzag moth because the lines on the wings of the female, which is white, run irregularly across them.

This scientist kept these imported moths at his house in Medford. One day he left a paper box containing eggs of this moth on the window sill. A storm came up and blew that box away. That one gust of wind is responsible for this great insect war, resulted in the death of millions of handsome trees, and the need of spending tens of millions of dollars to carry on the fight.

The Medford scientist knew of the danger in releasing gipsy moths in the United States. When he found that his box had been blown away he ran wildly about, taking every precaution he could to destroy the released prisoners.

It seemed at first that no harm had been done. In fact, it was twenty years before the gipsy moths appeared in such numbers as to attract attention. After that they came in such quantities that, in certain sections, the trees were stripped of foliage by their caterpillars, villages were overrun, sidewalks were slippery with their crushed bodies, the very dinner tables of the people were invaded. It was then that the resistance began.

The moth, it will be remembered, is a cousin of the butterfly. Both are lepidoptera, which means that they are insects which have scales on their wings. Butterflies hold their wings out straight, while the moths fold them and lay them down on their backs. The butterfly goes out in the daytime and the moth goes out at night. Otherwise they are much alike. Both lay eggs which hatch out into caterpillars which pass through the chrysalis stage and eventually become butterflies or moths.

There are native families of moths which have always lived in America. Everybody has seen nests of some of these spun on the tips of tree branches, there to harbor many caterpillars. Nearly everybody has had an opportunity to observe damage done to trees by these caterpillars. They eat the leaves; the trees become sickly and sometimes die.

But these native American moths rarely do great damage. This is because they have enemies in America that feed upon them and keep the supply of them within bounds. There is a balance of nature in so far as they are concerned.

But when these gipsy moths were brought in from Europe their enemies did not come along. As a result they developed more rapidly than they should have developed. Nature was thrown out of balance. The forests were threatened with destruction.

To know how to fight these gipsy moths the scientists had to know all their family secrets. They do not, for instance, spin their webs in the tree tops. Instead they lay their eggs in clusters of some 400 in cavities, under loose bark or stones, and cover them with a sort of hair. These eggs do not hatch until the following spring. Then they become caterpillars and crawl up the trees and eat the tender young leaves. After they have grown large and fat, they go into the chrysalis stage, hang themselves by a silken thread from a limb, and finally hatch into moths.

There are two or three odd things about this life cycle. In the first place the white mother moth, while she has well appearing wings, cannot fly. Her body is too heavy. The little brown father, however, can travel freely on the wing.

The fact that the mother could not fly raised the question of how the gipsy moth spread so rapidly over New England. The creature had been studied for thirty years in this country before the manner in which it spread was found out.

Strange as this may seem, it is a fact that the gipsy moth caterpillars can fly. They have no wings, yet they fly for miles at a time. Here is the way it comes about. While they are still small and light they grow an abundance of fluffy hair. At the root of each of these hairs is a tiny gas bag. Its business is to help them to float in the air. They are thus armed with hundreds of circus

balloons and they are so light that these balloons almost carry them away. Then they crawl out on the tip of a branch, spin themselves a bit of rope, and hang there until a wind comes along that is strong enough to break the rope. Such a wind may carry them distances up to a mile. So are the gipsy moths broadcast.

In this war on the gipsy moth there are certain steps that man himself

CATERPILLAR OF THE GIPSY MOTH.

may take to destroy them. He may put creosote on the egg clusters. He may put sticky bands about the trunks of the trees which will keep the caterpillars from getting up to eat the leaves, and they will starve to death. He may spray the trees with liquid poison, and the caterpillars will eat it and die. All of these are, however, very tedious and are regarded as mere aids in keeping numbers down. It would never be possible to reach all the moths in this way and actually get rid of them.

The real campaign must be based upon mustering in armies of other insects to fight the moths. It has been these other insects and not man that have been meeting the moths in the open and battling to the death.

The scientists tell us that about half of the insects in the world live upon plants. The gipsy moth belongs to this half. The other half live upon other animals or upon one another. They live largely upon the insects that live upon plants.

Of these insects that live on other live creatures there are two kinds. There are the predatory insects, which eat other creatures usually much smaller than themselves, just as the ladybird eats the scale insects, or just as the cat preys on the mice. Then there are the parasites, which are tiny creatures which live on larger ones, as the flea lives on the dog or as the tape worm lives inside man.

Certain birds are predatory as far as the gipsy moth is concerned. They attack it and eat it up. There are certain beetles, also, that are predatory. They devour great numbers of caterpillars. But it is the parasites that are the greatest enemies of the moths. The government and the State of Massachusetts have given more thought in this war to enlisting parasite armies than upon anything else.

But there is one predatory insect, not a parasite, which devours huge numbers of gipsy moth caterpillars. It is a glorious, shiny, green beetle named Calosoma which was brought from Europe for service against the gipsy moth. The American warlike beetles had failed their country in this emergency because they were satisfied with plodding along on the ground. What was needed was a tree-climbing beetle, and this immigrant from Europe was as agile among the branches as a Brazilian monkey. No caterpillar could go further out on a swinging branch than he. No height was so dizzy that he would not there clamp his victim and proceed to devour it.

Quite different from the dainty little ladybirds was this strapping, warlike cousin, more than an inch in length, a very tiger in its ferocity. It begins life as an egg laid on the ground in early summer. It hatches and comes out a grub, dominated by one instinct, hunger. Even the grub is a tree-climber and caterpillar-eater. It seizes the first fuzzy leaf eater it finds, cuts into it with its pincers, and begins breakfasting. At lunch time it may happen on a gipsy moth pupa, all sewed up in its cocoon, taking its nap before waking up a moth. It rips open this bag and eats the pupa alive.

These larvæ go on and on for two weeks with but one thought, that of stowing away food. Then they crawl into the ground and themselves become pupæ, fat fellows an inch long and half an inch thick. In ten days they are beetles. Oddly they have no appetites. Lazy fellows, they bask in the sun for a few days and then dig themselves in again in the ground and go to sleep. And a long sleep it is, for no alarm clock of Nature will wake them up until ten months have passed by.

Then there emerges Calosoma in all his glory. There seems to weigh heavily on his mind one fact—it has been ten months since he has eaten. The caterpillar season is at its height and he begins the slaughter. In one single day he will eat ten times his weight in caterpillars, and the next day he will go out and repeat the performance. This second summer of his existence he will lend his appetite to the destruction of caterpillars for two months. Then, despite the fact it is but August, he will again retire for a ten-month sleep, at the end of which time he will awake for another season of feasting.

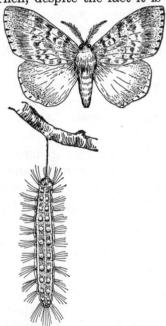

In his lifetime of two years he is likely to devour some 650 caterpillars. His is a considerable contribution to the re-establishment of the balance of nature in New England. But it took the gipsy moth 20 years to develop into great numbers and spread over New England, and the tree-climbing beetle must be given time.

Meantime the tiny parasites are at work. So complicated is this scheme of insect life that there are not merely parasites of the gipsy moth, but there are parasites of each stage of its development. There is, for instance, a distinct set of parasites that devote themselves entirely to its eggs.

THE GIPSY MOTH AND LARVA.

There is a tiny wasp called Anastatus which goes about looking for the egg clusters of the gipsy moth. Having found one of these, it pierces each egg, tiny things not bigger than the dot over an "i" on this page, and lays its own much smaller egg inside its shell. There the baby Anastatus hatches out and begins eating. Eventually it consumes all the contents of that egg shell, contents that the mother moth intended should hatch out into caterpillars. The invader grows until it fills the shell. Then it goes to sleep and waits until the following spring, when it emerges as a wasp to lay other eggs in bigger eggs. So does it help keep down the number of caterpillars.

There is yet another similar insect which, after many unsuccessful attempts, was brought over from Japan and established in New England. This Japanese midget, also a wasp, waits until the young caterpillar is developing in the egg, and then lays its own. It prefers a young caterpillar to an egg as food. This Japanese infant wakes up in the midst of food in the snug little house it has stolen and begins eating. If lucky it emerges to wasphood.

This method of pirating the moth egg as a nursery, however, does not always work out so happily. Here was a tragedy which the scientists with their microscopes one day observed.

Anastatus had laid its egg in the moth egg. The little one had hatched and eaten all the food supply that it found so handily provided. Then it had gone to sleep to dream of the coming spring when it should take to wing and fly about the great, wide, beautiful world.

And while it slept there appeared a jealous rival, this tiny wasp from Japan. It also raised its young in borrowed eggs. It usually expected to find little caterpillars in them, but it was as well satisfied if they were occupied with the children of Anastatus. Anastatus was likewise satisfactory food. So the Japanese laid its eggs just the same.

But to complicate the situation, two other Japanese wasps came along and likewise laid eggs in this same shell. The three of them hatched. First they fell to and devoured young Anastatus. Then there were three of them in the shell, three grubs of the same family. It seemed to be a part of their code, possibly because they were Japanese and therefore fatalists, that only one should emerge. So they began eating each other until only the strongest survived. Finally that individual stood forth like a gladiator with his foot on the neck of his fallen foe.

Another deadly Japanese wasp was brought to the United States to take part in the war on the gipsy moth. Its part of the campaign is to attack the caterpillars after they are feeding on the leaves of the tree. It lays its eggs in the living body of the caterpillers. There they hatch out and begin to eat and to grow. When they are two or three weeks old, they begin to wriggle out of the body of the caterpillar. Sometimes as many as a hundred eggs are laid in the body of one caterpiller. As they crawl out a little at a time they look like so many great warts growing out of the caterpillar's sides. Finally they come entirely out. They spin themselves little cocoons in which to slumber while they are turning to wasps. The wounded caterpillar dies in their midst, and with the cocoons clustered about it, gives the impression of a mother and its little ones rather than a victim in the midst of its sleeping slayers.

Yet another enemy of the gipsy moth, a two-winged

fly, not a member of the four-winged wasp family as were those others, makes its attacks in quite a different way. It lays very great numbers of eggs on the leaves that the caterpillars are eating. These eggs are so small that they are not noticed by it and are swallowed whole. Inside the caterpillar they hatch, but do not become active until it has become a pupa and hung itself out to be transformed into a moth. During this period the little larvæ within it begin to grow and entirely consume it. When they have done eating, they drop to the ground, there to hide until the following spring.

This war of insect on insect seems somewhat appalling in its frightfulness—seems unthinkably cruel. It would indeed be so if insects were capable of suffering pain as are human beings. The scientists say, however, that they are so constituted that they suffer no pain. Since the majority of them are born to be devoured, Nature has so made them that they can play their rôles without pain. The caterpillar, with its body full of eating grubs, goes on quite comfortably, and it would be a mistake to think of it as suffering as would a human being under the same circumstances. So is it shown that these insect wars are without the horrors to those engaged in them that are present in the case of wars between human beings.

Scores of these insects that are enemies of the gipsy moth have been introduced and established to a degree in New England. They are gaining ground and are expected in the end to re-establish the balance in such a way that the gipsy moth (and the browntail moth which has had a similar history in the same field) will cease to be harmful.

The United States has learned its lesson from several

sad experiences in the introduction of foreign insects, and now watches its borders very carefully to prevent the bringing in of any insect that might cause trouble.

Not many people ever stop to think of the value to the community of different odd bits of scientific information —of such facts as that of the existence of the tree-climbing beetle in Europe, or the Japanese wasp at the other side of the world as enemies of this invading moth. Fortunately there are men here and there who devote themselves to such studies, and who, when an emergency presents itself, have at least a basis of an understanding of what may be done. It has come to pass, possibly because we are a big nation and much exposed, that the United States Government maintains the largest and most highly developed organization in all the world for the study of insects. This is the Bureau of Entomology of the Department of Agriculture. Its business is to protect crops from hurtful insects. Some of the States also, as, in this case, Massachusetts, have lesser bureaus to study their particular problems.

QUESTIONS

1. You know the old saying "Mighty oaks spring from tiny acorns." What dreadful calamity befell our forest life as the result of a seemingly little accident?

    In telling the story keep in mind the following points:
    (a) The purpose of the experiment which the Medford scientist made.
    (b) His realization of the harm done.
    (c) The havoc wrought by the gipsy moths.
    (d) The money spent by our government in waging war on this pest.
2. The monarch butterfly and the gipsy moth both belong to the order *Lepidoptera*. What are their chief differences and their points of resemblance?

3. So far in our readings we have had some splendid illustrations showing how the balance of nature in the plant and animal world has been kept and broken. List all the examples you remember.

4. (a) Scientists are like doctors. They must know all the symptoms of the case before they can bring about a cure. What family secrets did they find out about the gipsy moth?

  (b) What procedure did they follow in battling with this enemy of tree life?

5. The war of insect on insect is Mother Nature's chief way of maintaining an equilibrium in her vast domain. How did man call in consultation all of the enemies of the gipsy moth? Take into consideration the contributions made by:

  (a) The predatory insect, the Calosoma—a tree-climbing beetle.

  (b) The parasites, i. e., the various wasps, such as Anastatus and the Japanese midget wasps that make their nursery in the moth egg, the Japanese midget wasps that attack the caterpillar when it feeds on the leaves, and the two-winged fly that attends to the pupa stage of the insect.

6. What important work does the Bureau of Entomology of the Department of Agriculture carry on for the citizens of the United States? If you are interested in insect life and its problems, some day you, too, may serve the United States in this field. Our big nation, so much exposed to insect pests, needs highly trained minds to cope with emergencies.

Chapter VIII

THE COCKROACH

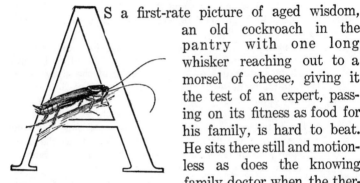

A S a first-rate picture of aged wisdom, an old cockroach in the pantry with one long whisker reaching out to a morsel of cheese, giving it the test of an expert, passing on its fitness as food for his family, is hard to beat. He sits there still and motionless as does the knowing family doctor when the thermometer is under your tongue and his finger on your pulse.

And he is as scientific. This instrument of his is as efficient as the doctor's thermometer. Place ever so little poison on that cheese, quantities that man with all his learning could not detect outside a laboratory, and the wise cockroach will immediately know that it is there.

The cockroach is to the insects what the crow is to the birds, what the fox is to the bigger animals. He is the creature of surprising wisdom, particularly in his relations to man.

Possibly this wisdom is because of his age. The cockroach is almost the oldest living member of the animal kingdom. By that I mean that in the form in which he exists today he has so existed in the world longer than any other living creature.

Through the millions of years on the earth conditions

have changed and with them animals and plants have changed. The horse was once a dog-like creature that ran wild about the plains, but as man learned how to make use of it, took better care of it, gave it better food, it developed into a much more splendid animal.

The great chrysanthemum as big across as a dinner plate was once an ordinary daisy of the field, but it was put into gardens, cared for, given favorable conditions under which to live, and so became the splendid flower of today. The dodo bird used to live along such icy shores as those of Labrador. It was a big, helpless, wingless creature that could not escape the modern huntsman, and so, when man became plentiful on those shores, all the dodos were killed. There is no longer a living dodo in all the world. Mastodons, those hugh creatures bigger than elephants, no longer exist. The buffalo came very near to being exterminated, while the cow, useful to man, has increased in numbers.

So have certain creatures come and gone. So have certain creatures grown stronger and others weaker. Conditions on the earth have changed, and with them the animals and plants have changed. Take, for instance, that age in which the world was warm and damp and was overrun with rank vegetation. It was one great jungle inhabited by huge reptiles. Naturally the creatures which lived in that sort of world would be different from the creatures that live in a world such as we have today.

What was once a great swamp might very slowly dry up and be transformed into a grassy plain. The creatures that lived in the swamp would have to so change that they could live on the plain or they would all die. Changing conditions have led to these changes in animals and plants throughout the development of the world.

It was the great deposits of the leaves and stalks of plants in the age of rank vegetation that created the coal deposits that we are now using. In this coal was left the imprints of the bodies of animals and insects that lived in those times. This photograph gallery of millions of years ago shows that there were then no flies, bees, wasps, butterflies. There was no man. There were almost none of the creatures that now exist.

But the cockroach was there. In those far distant times it looked very much as it does today. It was, in fact, very abundant. It liked the warm, damp weather and the plentiful food supply. It was the heyday of the cockroach.

But of all the creatures that lived in that age the cockroach is almost the only one that has come down to modern times looking anything like it did then. Others have had to change to meet new conditions that have developed, but the cockroach, if you please, found itself well able to get along under any circumstances. It could not have done this, of course, if it had not been a pretty fit sort of person. It was as though the maker of an automobile had put out a design that was so good that he did not have to change it as the years passed. The design of this cockroach, developed millions of years ago, was so good that it has not been necessary to make any changes.

Compared with the cockroach, man is a new thing in the world, an experiment that is just now being tried. He has not yet shown his fitness to survive. Perhaps he will fail. But the cockroach! Well, there is the record of the ages.

The cockroach is a creature of the dark, of narrow spaces, and of warm places. It still clings to some elements of that nature which made it so successful in that age of warmth so long ago. There are wild cockroaches, but those kinds we know best live chiefly in the houses of man because they are warm and because there is food. They are completely domesticated. Of all man-built establishments in colder climes cockroaches like bake shops best. Next to these come kitchens and pantries which combine warmth and food supply.

Primarily, however, the cockroach is a creature of the tropics. It thrives in great abundance in warm countries, grows much larger, is much more of a pest there than in temperate regions.

Life aboard ship appeals particularly to it. The darkness beneath decks, the innumerable opportunities to hide, the warmth, are to its liking. Few creatures are better fitted to play the rôle of the stowaway and ride all about the world on ships than is the cockroach. It has taken advantage of this fitness since the time ships began crossing the seas. Many varieties of cockroaches were not known outside the tropics until ships began threading the oceans.

In England the cockroach is popularly known as the black beetle despite the fact that it is brown and is not a beetle at all. It did not exist in England, however, until some three hundred years ago. It is believed that it came from India in those days when ships first began going to that part of the world for spices. There is a record of a certain Spanish ship captured by Sir Francis Drake, famed buccaneer of his day, a ship laden with spices from the East. It is set down that this ship contained innumerable strange insects hitherto unknown to the English, but descriptions of which indicate that they were undoubtedly cockroaches. Even the name cockroach is traced back to the Spanish, who called these insects "cucarache," which, strictly speaking, is a woodlouse. The records show them appearing first in the port towns and spreading thence inland. So ships have carried them all over the world.

There are roaches in practically every house. The careful housekeeper is likely to flatter herself that these marauders do not exist in her spotless kitchen, but if she should steal in silently in the dead of night and switch on a light she would probably see many of them scamper to cover.

In very many households they exist in great numbers. Waste food is their chief encouragement. They will eat anything that man eats and man is given to an unusual range of food. Cockroaches would have little sympathy with the ladybird which must have one special morsel of food or nothing. They have been known to eat the tops of shoes, the blacking that goes on them, the paper off the wall, woolen clothing, their own dead. In the tropics they infest the beds of their sleeping hosts. On infested ships sailors are sometimes forced to sleep with mittens and socks on because of the tendency of cockroaches to eat their fingernails and toenails, delicacies for which they show a special fondness. They like ink, while their fondness for stale beer is beyond measure.

When the light flashes on, they scurry to their hiding places behind moulding boards, into cracks, beneath boxes. They are built for hiding in narrow places, being flat and thin. They run with great speed on their six sturdy legs. An odd thing about these legs is the fact that if one of them is broken off (not too short) another will grow in its place.

There are two or three very strange things about the form and habits of the cockroach. There is the matter of the eggs, for instance. The cockroach is a wise and economical sort of creature and does not approve of the wastefulness of younger generations of insects in laying hundreds of eggs to raise one child. The cockroach lays few eggs, but takes good care of them. Certain of them, for instance, lay just sixteen eggs. These are carefully arranged in a little satchel, puckered at the top where the heads of the young ones are intended to emerge. This satchel is carried about by the mother until the eggs are nearly ready to hatch. Cockroaches have been seen moving from one house to another, each mother carrying her basket of eggs.

When they are ready to hatch, the mother tucks the bag away in a safe and narrow crack. She may even go so far as to scratch open the fastenings at the top of the bag. The babies come out. They are tiny creatures not bigger than pinheads. The mother thereupon gives them an example in economy by eating up the satchel in which the eggs were carried.

Members of this straight wing order of insects do not go through a larval and pupal stage as do the scale wings. When they come out of the egg they are of the general form of their parents and have nothing to do but to grow to become marauding cockroaches. Among cockroaches, however, there is more home life than among most insects. They are a sociable lot. They stick together and are given to imitations of such games of man as riding piggy back.

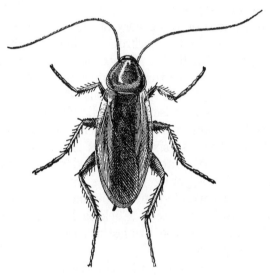

THE COCKROACH.

Each cockroach has two antennæ, long, slender, delicate whiskers, or feelers. The antennæ of the cockroach (not that it is the only insect that has them) are the most remarkable instruments carried by any living creature. Certain insects have certain senses very highly developed. There is no question, for example, but some of them can see much better in the dark than can man.

A butterfly has a very delicate sense of smell and can locate its mate by that sense at greater distances than by sight.

These antennæ are instruments of very delicate sensitiveness. They begin in the head with a first-class ball-and-socket joint, which turns each way, and they are often long enough to reach the full length of the body and meet beyond the tail. There may be 100 joints in them.

The microscope shows that there are little dots on these antennæ toward the end. These are sensitive spots connected with the brain of the insect. Some of these dots are sensitive to odors. Thus does the antennæ become the nose of the insect. Place a piece of food near a cockroach and one of these antennæ immediately swings around to it, sniffs it, reports back.

The cockroach feels through its antennæ. There are sensitive spots for this purpose also. And it hears through them. They are its ears. Through them it feels very delicate vibrations, and sound is a matter of vibration.

There are those who maintain that the plan of placing the nose and the ears at the end of a slim, flexible member like the antenna is a much better one than that of placing them on the head as in man. If man's nose and ears were thus attached to a whip-like member longer than his body he might more easily smell a rose while walking in the garden or get his ears to the ground to detect the approach of his enemy.

It is largely because of their antennæ that cockroaches are so clever at outwitting man—partly this and partly through having lived so long with him that they know all his tricks. Those cockroaches that learned to outwit man have survived, while those that did not learn to do so have been killed off, have ceased to exist. So has the law of the survival of the fittest worked in the pantry as elsewhere.

In nothing does the cockroach show its wisdom better than in its refusal to allow itself to be poisoned by man. The pantry may be full of them, other food may be kept away from them until they are ravenous with hunger. Then a palatable, damp mixture of dough, just to their liking may be prepared and only the tiniest bit of poison stirred into it. But the cockroach will not touch the mixture. In France the children say, "My little finger tells me." In the case of the cockroach it is probably its little antennæ which tell it of the poison.

Almost the only successful way to get rid of cockroaches is to put insect powder in their runways and hiding places. They do not eat this powder, but breathe it into their bodies.

This brings out another very interesting fact with relation to the construction of insects—the fact that they have no lungs. In the case of the cockroach, for instance, it has a series of openings along the lower edge of its body. It takes in air through these openings. In our bodies we breathe the air into our lungs and the heart pumps the blood into them. The blood and the air are brought together and the blood is purified by the air. Instead of bringing the blood to the lungs to be purified, the insect sends the air all through its body and it purifies the blood where it finds it. There are no lungs, but a system of air tubes.

The blood of these insects, it here appears, is a quite

different thing from the blood of other than the jointed creatures of the animal kingdom. It is different primarily because of the fact that it is not red. Insect blood and that of their jointed relatives, such as crabs and spiders, is pale and almost colorless. It is none the less blood. It performs the same functions as does the blood of other animals, yet in appearance it is not the sort of liquid one is accustomed to style blood.

But this way of breathing makes it hard for the insects when poison powder is sprinkled about their haunts. It gets into their ventilator systems and goes all through them. It kills them. Scientists have learned to get rid of insects through an understanding of these queer facts with relation to them. One might not at first appreciate the benefit that might come to man from a careful study of the manner in which insects breathe. In this complicated world, however, it usually turns out that an opportunity will somewhere be found to use any odd bit of information that scientists may develop. So it happened that the most effective insect powders were developed upon an understanding of the strange way in which insects breathe.

There is general agreement that the cockroach is a first-class nuisance. It eats up great quantities of good food. It pretty well messes up the place. It spoils much food that it does not destroy by giving it a peculiar odor and flavor. The cockroach has glands on its sides that secrete an ill-smelling fluid. If it gets into the sideboard and runs about over the dishes it leaves some of that odor on them. The fact that a cockroach has crawled over a soup dish may give the soup afterward served in it a very bad taste. It is likely to give the same taste to any food that it touches. So does it become disagreeable and so do housekeepers resort to insect powder.

But occasionally it performs an act of helpfulness for man. It is, under certain conditions, a great slayer of bedbugs. These creatures are also hiders in narrow places. They live in cracks, under the wall paper, among the bed springs. The cockroach knows their haunts and is so constructed that it can get into them. It will not go hunting for them, however, if there is plenty of food in the pantry. If that food supply is shut off, it will fare afield for live game. It might easily happen that the family whose house was inhabited by both of these creatures would go away for a vacation, thus failing to maintain the pantry stock of cockroach food, and upon returning, would find that their supply of bedbugs had disappeared. The hungry cockroaches would have eaten them.

Despite this incidental service the cockroach must be definitely set down as an enemy of man, one against which it will be necessary for him to maintain constant opposition.

The cockroach is a member of one of the most interesting groups of insects in the world, six of them, all by way of being familiar acquaintances of most of us. Of these cousins the grasshopper, the cricket, and the katydid sit on one side, and the cockroach, the walking sticks, and the devil-horse on the other. The first three are jumpers and the latter three are runners. An odd thing is that these three jumpers are also musicians and that the runners are not. This does not necessarily prove that there is any connection between jumping and music. There are but four insect musicians in all the world, however, and three of them are these jumpers.

The wings of all these insects fold down their backs on the principle upon which a fan closes, in straight lines. For this reason they are classified by the scientists as "straight wings," and are given the Greek name, Orthoptera, which means "straight wings."

QUESTIONS

1. "The cockroach is to the insects what the crow is to the birds, what the fox is to bigger animals." Give illustrations proving this.
2. A close study of plants and animals reveals many interesting changes or adaptations to environment made by them during the course of the ages. What new things have you learned about the horse, the chrysanthemum, the dodo bird, the mastodon, illustrating these laws?
3. How do we know that the cockroach is one of the oldest members of the animal kingdom? What proof have scientists to offer when they maintain that the cockroach changed its appearances very little during a million years?
4. (a) The cockroach is primarily a creature of the tropics. How do you account for its spreading to a more temperate climate? What historical reference has been made to this pest?
   (b) What places do cockroaches select as their homes?
   (c) The ladybird, you may remember, was most particular about her diet. Compare the menu of the cockroach with that of the ladybird.
   (d) Many insects we have considered thus far have made sure that they would have young ones by laying hundreds of eggs. The cockroach practices many civic virtues. In what respects do we see thrift observed and sociability upheld?
5. (a) How do antennæ serve the cockroach as gateways of knowledge?
   (b) How does the cockroach help the housekeeper?
   (c) Why is the cockroach so unpopular with the housekeeper?
   (d) How does the wise housekeeper exterminate this nuisance?
6. Using your own specimens as models, make drawings of the runners and jumpers in the insect world so far considered.

CHAPTER IX

THE CICADA

TAKING it all in all, the cicada is probably the most remarkable individual in all the insect world.

It is, for instance, the noisiest of these little jointed, six-legged creatures that wear their skeletons on the outside and constitute such an important branch of the animal kingdom.

Despite the fact that it holds the insect record for clamor, it may be stone deaf and never hears even an echo of all the racket that it makes.

It is unusual in almost any respect from which you see fit to examine it. Almost nobody in America, for instance, knows it by its proper name. It is called "the locust" when it is not a locust at all. The short-horned grasshopper is the locust, has been known as such all around the world since Biblical times, and the cicada is not even a relative of the grasshopper family. But the early settlers in the United States got to calling it "the locust" and now it is generally known by that name. There is no way to stop people calling a cicada a locust and they will probably go on doing so. But when one writes a book he is bound to use the right name.

The cicada is also called the dog-day harvest fly, the lyreman, the seventeen-year locust. The seventeen-year kind appears about the middle of May, sometimes in dense swarms, and stays until the Fourth of July. It is

a big, blocky, sturdy insect with none other in the world
that at all resembles it. It is a thing apart. The harvest
fly comes in the late summer.

The cicada has a most alluring life history; it is the
Methuselah of its kind. But more of this later. Let us
begin with its noise and its deafness.

It is most remarkable how this insect specializes on
noise making, while all others, with the single exception
of the grasshopper family, are born to lives of silence.
It is specially built for noise making. It is a living sounding board. If a man could fit himself out with a noise-
maker as effective in proportion to his size as that of the
cicada, he could single-handed drown the noise of boiler
factories and machine shops.

The cicada accomplishes these results by means of
drums. It has two such under its abdomen, made of
dry, crisp, ribbed membranes. These drums are cupped
out and look something like shells from the sea-shore.
To the inside of them are attached strong muscles. With
these muscles it vibrates the stiff membrane in and out.
It is the same principle that causes a sheet of somewhat
cupped tin or the bottom of the dishpan to rattle·noisily
if it is pushed back and forth. But the cicada holds the
record for getting the greatest amount of noise out of
the smallest drum.

This insect has two or three distinct calls. They may
not be as plain as the call of that member of the grasshopper family which lives in the trees and contradicts
herself by insisting that "Katy did. Katy didn't." In
one of its calls the cicada is quoted as saying "Pha-r-
R-R-r-ach," with emphasis at the middle and quite

drawn out.   Scientists have tried to express another of its calls in letters.   It spells out in this way: "tsh-e-e-E-E-E-E-e-ou," likewise emphasized in the middle and likewise long drawn out.

The male insect makes the noise while sitting on a limb with his sweetheart.   He keeps it up from sunrise to sunset, but, fortunately, is silent by night.   At his best he appears at the same time as millions of his fellows and the noise of the lot of them becomes a veritable din.

A French scientist became curious to know why the cicada sang and if he liked his own music.   He had observed that the cicada was shy, had very excellent pop eyes, and would stop singing and fly away when he saw a man creature coming.   He noted, however, that one could approach him from behind, out of the range of his eyes, and he would not fly.   From this position the scientist clapped his hands, cried out, knocked stones together.   The cicada paid no attention.   The scientist lived near a small French town at which were kept certain cannon for firing salutes.   He borrowed these cannon, loaded them, slipped up behind a singing cicada, and fired them.   That insect never missed a note of his song.   The scientist decided that making noise was the cicada's instinct, but that it had no sense of hearing.   American entomologists take issue with this conclusion, however, arguing that the sense of hearing of the insect might be attuned to respond to sounds in one key, but not to sounds in another.   The cicada might hear its own song and yet not hear the cannon.

The cicada belongs to the *Homoptera*, which word, taken apart, means "same wing."   It would have been more descriptive of one of the basic traits of the group if it had been given a Greek name which means "sap sucker." It belongs to that group of insects which lives by drinking the sap of trees and shrubs.

Tiny scale insects that can hardly be seen with the naked eye make up the greater part of this group. They are near relatives to this giant cicada which is one of the largest of insects. All have augers which they sink into the bark and through which they drink sap. It was the cottony-cushion scale of this family that once threatened all the orange groves of California until the ladybird beetle was introduced and ate it up. Though the scale insect may be no bigger than the point of a pin, the cicada is an inch and a half long when at rest and may measure three inches from tip to tip of its wings when in flight.

The dainty little ladybird beetle makes it her business to eat up scale insects, but she would have a hard time if she tackled this burly cicada.

All of these sap suckers are enemies of man. If they were not restrained they would drink so much of the sap from plants that many of the latter would die and their kind would cease to exist. The plants useful to man are especially liable to attack. The entomologists say that if it were not for the enemies of the sap suckers, the other creatures that prey upon them, they would spoil the world in half a year. Man could not live on the earth if these creatures were not kept down.

Again, it is queer that these cicadas, giants of their kind, are among the least injurious of the sap suckers. The tiny scale insect is much more deadly. The cicada, however, does its share of damage.

The oddest of the cicadas is the periodical cicada which has come to be improperly known as the seventeen-year locust, and which challenges them all in the strangeness of its life cycle.

On a July day its egg hatches on the limb of a tree. The tiny creature that emerges is queer and fish-like in appearance, but remains so only until it has dried out a bit, when it sheds its skin and takes on a new form. This skin attaches one end of itself to the limb of the tree, while the newly emerged little creature, no bigger than a gnat, hangs by the other end and gives itself the treat of a sun bath. It tries out its legs, exercises a most desperate looking pair of claws attached to those in front, waits for its body covering to harden. This done, it turns loose and takes its plunge into the unknown. Fortunately it has acquired a crust, for it may be quite a distance to the ground and there may be a jolt in landing.

And this bit of a swing in the sun is the only peep that this little creature is to have at the great outside world for seventeen long years. When it strikes the ground it begins looking for a crack into which it may crawl, finds it, plunges in. In doing so it banishes itself to darkness for nearly two decades. Not only this, but it plunges into solitude. For the next seventeen years it will remain in a solitary dungeon, will not even see another creature of its kind, will scarcely move in its narrow cell. Then will come its day of freedom and glory before it breeds another generation to do the same, and dies.

This life of the cicada underground is something of a mystery. There is little opportunity to observe it. Different families of cicadas stay under different lengths of time. The European scientists incline to the belief that the cicada over there stays down four years. There are some that emerge each season, but there are also some that are going down each season. In the United States the surprising fact that most of them stay under for seventeen years is unquestionably established. There are cicadas every year, however, but whether these are kinds that stay in the ground only a year or whether they are but chance, irregular members of the big broods is not definitely known. The dog-day harvest fly which comes out late in the summer undoubtedly makes the round trip of its existence in a single year. The detailed account contained in this book has to do with the periodical cicada.

There are great areas covering whole states where the cicadas appear once in overwhelming numbers, disappear, and are not seen again for seventeen years. Then the multitude reappears, has its noisy fling, and again is gone. In seventeen years more it reappears, repeats its former performance, and is again gone. Each seventeen years it reappears.

While one brood is making these regular but infrequent visits, another nearby is doing the same thing, but coming to the surface in different years. Thus does it happen that the different tribes never see each other.

There are broods in America of which records have now been kept for nearly 300 years. There was one near Plymouth of which an observer wrote in 1633 that:

"There was a numerous company of flies, which were like for bigness unto wasps or bumblebees, they came out of little holes in the ground, and did eat up the green things, and made such a constant yelling noise as made all the woods ring of them, and ready to deaf the hearers."

This brood has appeared every seventeen years since that time. The entomologists have studied this and other broods. There are in all about thirty broods in the country. Each is numbered and has its records of appearances. Each goes on its uninterrupted cycle and there is no mixing of broods, for no two come to maturity at the same time. The southern broods stay down but thirteen years, the development being faster where it is warmer. Sometimes it happens that the seventeen-year and the thirteen-year cicada overlap and come out at the same time, but even then they refuse to associate with one another.

The greatest of all the broods is Brood X. There is a clear record of this brood since 1715, but in 1902 and in 1919 the periodical reappearances of these insects were very carefully observed. They appear at the same moment over a territory that covers half the area east of the Mississippi. There is one great group of which Delaware is a center, another grouped about Indiana, and another straggling from Virginia to Georgia across Tennessee. The recurring appearances of these clamorous insects has come to be an event of which much is made in many parts of the United States.

When the baby cicada dropped from its limb and crawled into the ground it knew what it was about. It thought that this ground beneath a tree contained roots. It burrowed about, and if possible, found one of these, sank its beak into it, and there established its home. If it found no root it might, in an emergency, live on some such thing as decaying vegetation. In hunting a root it demonstrated its kinship to the scale insect. It lived on the sap of this root. It was content to stay right there and drink sap. As a matter of fact, it did stay right there for those seventeen years.

There were likely to be other cicada grubs nearby, likewise attached to food-producing roots. But they never visited about with each other. Each one kept to itself. Winters followed summers, world events transpired, automobiles came into use, airplanes were developed, wars were fought, between the appearances of these cicadas. Below ground there was not much evidence of this passing of time. There were no clocks, no calendars hanging on the wall, no sunrises and sunsets, little changing of seasons.

Yet inside the cicada there was some mechanism which counted the passing of time. Finally a day arrived which was some three months less than seventeen years. The cicadas down by Oyster Bay knew the day had come. So did those in Georgia, in Maryland, in Michigan, along the Wabash. Each one of them in all that area inhabited by Brood X removed its beak from the root that had so long nourished it on very near the self-same day, gave up its inaction, and entered another phase of its career. It began a climb to the light of day which it had seen so briefly seventeen years before.

This lusty grub may have been attached to a root a foot beneath the ground. It starts up. In doing so it excavates a tunnel nearly an inch across and a foot long. After it has come out the tunnel is there and anybody

who wants to may examine it.   Many scientific men have
examined such tunnels.   The insect has made a journey
of a foot and left this track.   It began at the bottom and
ended at the top.   Where, they have wondered, did it
put the dirt that it took out of the tunnel?   There was
none of it piled at the mouth of the tunnel.   Where did
it go?

This was a puzzle until, finally, somebody put some of
the grubs in glass tubes, filled them with dirt, and watched
them fighting their way upward.

They are sturdy insects with most remarkable front
legs which they can use as pick, rake, and tamping sledge.
The pick digs down some loose dirt from the roof of the
tunnel.   The rake gathers it up, mixes it with moisture
from inside the insect which it evidently replenishes from
the sap-bearing root.   This damp earth is gathered up in
the elbow of the front leg.   Then, with surprising force,
it is driven into the wall of the tunnel.   Then it is tamped,
first with one front leg and then with the other.   The grub
pounds at it like an athlete exercising at a punching bag.

It is thus that the tunnel is made without leaving rubbish.   The earth all around is tamped down until it occupies less space.   This space becomes the tunnel.

This tunnel comes very near to the surface of the earth,
within a quarter or an eighth of an inch.   Under certain
conditions, most notably when the ground is wet, the
cicadas build towers of mud for themselves four or five
inches above the surface.   These may often be observed
in swampish places.   But in towers or not, having got this
near to the light of day, they seem to wait for some zero
hour of nature, for some time signal for breaking through.

It is likely to come at sunset, May twenty-first. Millions of maturing cicadas that evening hear the call inaudible to man with all his cleverness, the summons for which they have waited so long. They burst the caps off their tunnels and scramble out. Millions of them, like soldiers from the trenches going over the top, appear at the same moment. The earth is alive with them. Two score of them may emerge in a single square foot of ground.

And once out, they are in a great hurry. They begin crawling as rapidly as ever they can. What they are looking for is a tree up which to climb. Available trees become literally covered with them. Failing to find a tree, they will accept a shrub, in an emergency even a blade of grass. But they must climb upward.

THE GRUB CHANGING INTO THE WINGED INSECT.

Finding the best place available, they dig in their claws, attach themselves quite securely, become very still. Then they proceed to that act of transformation that is common to all insects. They hunch their backs, split their skins down the middle. A strange and different form begins to wriggle inside, and gradually there emerges the glorious, winged cicada in its final form. Bright yellow is its body and wings, with spots of black near the head and brilliant red eyes. Its body is destined to turn black soon, its wings amber and its eyes to remain red. It is less than beautiful, but very distinctive, very individual.

Then on the morrow at sunrise the song breaks out. It becomes a roar, for the trees are already full of the insects. It increases day by day. Each cicada soon finds its mate and the two of them choose a place in the sun and idle away all the days of a glorious June, to the unending clamor of a perhaps unheard but never ceasing song of the male. Each pair of cicadas in all the wood does just the same thing and the din of it never ceases.

While the male sings, the female busies herself with cutting slits in the branches of the trees and laying her eggs within them. The sap that so many cicadas drain from the trees is quite weakening to them, but these wounds, made by the female in making places to deposit her eggs, are much more serious. In young fruit trees they sometimes prove fatal. They weaken many branches that later break in the wind.

The periodical appearance of the cicada in such stupendous numbers has always been a frightening event. The Indians and the early settlers thought that they brought plague. Farmers, knowing of the harm done by the grasshoppers, have regarded the cicada as a similar menace. It is true that they do no little harm to forests, orchards, and nursery stock, but it is likewise true that they do not carry ruin in their wake.

From May 21 to July 4 is the normal span of the life of the periodical cicada in the winged stage. As June draws to an end the grim reaper, in the form of old age, cuts down these multitudes of clamberers. Their ranks thin. They steadily disappear. A patriarch here and there may raise his voice toward the middle of July. But soon there comes a day when there is not a living cicada of these strange recurring broods above the ground in all the world. The tiny grovelers are again below the ground searching out roots to which to attach themselves for the long gloom, but above the surface there is but the occasional dead skin clinging to a shrub where once the trumpeter changed from an ugly nymph and came into its glory.

True to the predictions of the scientists, Brood X of the cicadas appeared in May, 1936. While scientists watched and photographed them by flashlight, they left their earthy burrows at sundown, climbed to the top of the tree whose roots had nourished them for so long, or to any bush on which they could get a toe hold. There they pulled and twisted themselves out of the casing which held them and emerged as white moths turning darker and darker until the following morning they were entirely black save for the orange-red W on their wings. Then began the noisy mating calls of the males that have caused superstitious people to fear them all these years. The females flew to find some succulent tips of trees where they might deposit several hundred white eggs, which in a few weeks became larvae that later dropped to the ground to begin all over again their existence as pupa until in 1953 nature will call them again to emerge into life.

QUESTIONS

1. (a) The Cicada may be regarded as a specialist in noise making. Do you think the poet, Elizabeth Akers, agrees with this view when she writes:

      "The shy cicada, whose noon voice rings
          So piercing shrill that it almost stings
              The sense of hearing."

   (b) Describe how Mother Nature equips the Cicada for noise making.

2. One of the ancient writers once wrote:

      "Happy the Cicadas' lives
          Since they all have voiceless wives."

   From this statement do you think the ancients had observed the habits of insects?

3. (a) The Cicada and cottony cushion scale belong to the same order. Why would the term "sap sucker" fit this group of insects?

   (b) Compare these two insects as to size and harmfulness.

4. (a) Why are some Cicadas called seventeen-year locusts?

   (b) What interesting observations—going back as far as the early seventeenth century—have been made in the United States concerning their periodic appearance?

5. (a) Describe the underground life of this hermit Cicada.

   (b) How does he make his tunnel?

   (c) What part of the story interested you most? Read this part again and be ready to tell it in your own words.

6. Compare the seventeen-year locust with the cottony-cushion scale.

7. If, on some balmy June night, you should chance upon a tree covered with cicadas you might think that it had suddenly blossomed into beautiful flowers in the moonlight. What effect would this produce upon the farmer? Why?

CHAPTER X

THE PRAYING MANTIS

ERHAPS you would like to know what is the real character of this insect which appears to be the most pious of all its kind, this creature of the leaves that now lifts its arms in benediction as does the preacher at the end of his sermon, and now bows itself upon its knees as though in prayer.

Its very name, mantis, means diviner, or fortune-teller. The English call it a soothsayer and old-fashioned people over there believe that its long finger will point the way home to a lost child. In France young women go to the cross-roads and ask the mantis from which way their lovers will come.

To the Hottentots, in Africa, these insects are gods. The Turks regard them as fellow worshippers. The negroes in our Southern States call them mule-killers, believing that their "molasses" will cause the death of work animals, but also that they can detect the presence of angels. They are also called devil-horses, rear-horses, camel crickets. They are creatures of mystery, reverence, fear. They are meek, slow-moving, with oddly alert faces. But whether they are saints or villains, few people actually know. I shall lay the evidence before you and let you draw your own conclusions.

In attempting to form an opinion of human beings, it is not an unusual proceeding to ask who are their relatives. Applying this test, it will be found that the mantis is of an order of insects of which you already know a good deal. It is one of the straight-wings, the

Orthoptera, and therefore a cousin to the grasshopper and the cockroach. The straight-wings, as you will remember, are divided into the jumpers and the runners. The mantis is a runner, but comes near being in the snail class, since it moves so slowly. Speed in running is not a part of its rôle.

It practices the same sort of deception as does the grasshopper and takes on a greenish color that it may hide in the leaves. In South America there is a sort of mantis that is given to brilliant colors. This is that it may hide from its enemies in the beautiful orchid clusters with which the trees abound.

The mantis has a hinge in the middle of its body and the part in front of it

THE MANTIS AT PRAYER.

has the appearance of a giraffe-like neck. Like all insects it has six legs. The hind legs are oddly developed in the grasshopper, the object being jumping, but in cousin mantis it is the front legs that are peculiar. They are very long, so big as to be quite out of proportion to the rest of its body, and have two well-developed elbows with barbs inside to help in gripping whatever they

THE PRAYING MANTIS

may seize.   It is these arms that constitute the business
end of the mantis.

When the hinge bends in the middle, the front part of
the mantis sits up straight.   Then an odd thing is discovered.   This mantis can turn its head about as can a
human being and look from side to side.   It is the only
insect in the world that can do this.   It almost has a face
with a very alert and wide-awake look upon it.

The grasshopper, the cricket, the katydid, the walking-
stick, all relatives of the mantis, eat plants.   The cockroach, living in the house with man, has learned to eat
the master's food.   But along comes this cousin and
refuses to have aught to do with such tame bills of fare.
It will eat, if you please, naught but raw meat.

If you watch this meek, slow-moving creature of religious mien on a shrub or, if you are in South America, on
an orchid, you will be likely to get one of the greatest
surprises of your life.   Solemnly and slowly it may
advance, like a minister in his pulpit, until it has reached
a point of vantage.   Then it will rear itself, by dint of
the hinges in its back, and lift its arms on high.   There
it will remain motionless as though in worship.   Presently a katydid may fly past.   Instantly there is a transformation.   With lightning swiftness the mantis will
strike with one of those poised arms.   It will reach out a
surprising distance.   When the arm comes back it will
have in its elbow—with no danger of getting away, because of the barbs—the struggling form of the katydid.
Kick and strike as that creature may, it cannot break
the clutch of the mantis' elbow.   Without more ado,
the latter sets about devouring the katydid.

Upon occasion it is even more spectacular than this. The mantis may sit there on its leaf very still, as though in prayer.   A grasshopper may approach innocently, little knowing the danger.   It is yet beyond the reach of the mantis.   Suddenly there is a transformation, a change in character that is surprisingly like that of Dr. Jekyll to Mr. Hyde.   The pious mantis rises on tiptoe, flings out its wings, straightens up its body, stretches out its dangerous arms.   It becomes a creature frightful to see, an

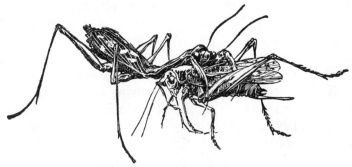

MANTIS SEIZING A GRASSHOPPER

ogre, a terror.   The grasshopper comes to a halt.   It stands as though spellbound.   It is fascinated, charmed. The drive in those hopping legs, which might so easily take it out of danger, seems paralyzed.   It may even approach the mantis or the mantis may slowly advance toward it.   Soon it is within the range of those vises.   A flash, and it is in the grip of the executioner.   The grasshopper has met a tragic death.

This slow-moving creature is by way of being the monarch of the insect world.   It is little respecter of persons and will tackle almost any other insect that comes its way. Bees, wasps, grasshoppers are its most frequent victims. Its appetite is most surprising. One observer saw a mantis in one forenoon devour three grasshoppers, a grandaddy-long-legs, and then, horrible as was the spectacle, fall upon one of its own kind and proceed to eat it.

The mantis is a cannibal. It would as quickly eat another mantis as some other creature. They are much given to fighting among themselves. In Java and in China mantis fights are staged by the natives and they place bets on the results as do the Cubans on cock fights. The victor gets its pay by being allowed to devour the vanquished foe.

The prime crime of them all in bloodthirstiness, however, is that committed by the female mantis on her mate. The females are larger and stronger than the males and, when the latter seem to have lost their usefulness and become unpopular about the place, the females eat them up.

This evidence pretty well does away with the theory that the mantis is a religious sort of person with any tendency to practise the Christian virtues. Yet there seems to be an element of sentiment in this harsh creature, a liking for the companionship of man and an understanding of his ways. The mantis may be tamed, may grow to be a pet. It adjusts itself, for instance, to being tied to the bed post by a silken thread. It learns to take its food from the hand of its master and may even be trained to stand guard for him against mosquitoes, snapping up any prowler that settles down upon his hand for a dinner of blood.

The belief of the negroes that the mantis is a "mule-killer" is based on the fact that it "makes molasses" or "spits tobacco juice," and this is supposed to be deadly. If the mantis is caught in the hand, for instance, a very dirty sort of liquid immediately begins to flow from its mouth. This liquid is not only messy in the extreme, but it has a bad odor. It is unpleasant to the captor. The mantis knows this. That is the reason it uses it. It hopes that, by making itself disagreeable, it may gain its release.

This means of defense is rather pathetic when used by the mantis against so big a creature as man. Man would not be likely to be turned from his purpose by so mild a weapon. In the case of the other enemies of the mantis this defense may be more effective. Insect enemies are small and this liquid may gum up their legs and antennæ and cause them much trouble. Insects also have much more delicate senses of smell than have human beings and the odor of this liquid may be quite impossible for them to endure. Thus it is an effective defense.

Other insects resort to this secretion of disagreeable liquids to drive their enemies away. Grasshoppers are notorious for the practice. The caterpillar of the milk-weed butterfly thus makes itself distasteful as a food for birds. Even the cricket, when captured, gives out a bad smelling liquid from its abdomen. The oil beetle goes to the extreme of shedding drops of blood from the joints of its legs. This blood has an acid in it that is so strong that it may even make blisters on the fingers.

Many insects have these scent glands and use them in attempts to repel their enemies. It is the same princi-ple as that employed by the skunk, but less highly developed. To be effective it does not need to be so highly developed, as the odor of a mantis is probably as offensive to another insect as the odor of a skunk is to its bigger animal enemies.

But aside from the rather mild unpleasantness of the liquid secreted by the mantis, it is entirely harmless. It certainly could have no effect upon that tough individual, the mule. Aside from this and a bit of pinching and clawing, the mantis has no power to injure whoever would capture it with his bare hand. It is a desperate fighter in the insect world, but capable of little resistance to larger creatures.

Anybody who has lived in the mantis country, all that region south of New Jersey, has seen its egg masses, though he may not have known what they were. Sticking about on twigs or tree trunks through the winter they may be found, an inch long, as big as a finger, yellow, and fluted somewhat to resemble heads of wheat. Down the middle of them are two rows of light colored eggs, so arranged as to give the appearance of a braid. It is an artistic, delicately sculptured creation, sticking there on the limb as a monument to a builder that is gone.

The making of this egg case is one of those remarkable examples of masonry in which the insect world abounds. It is made by the mother mantis, her mate not being there to do any of the work, since he was devoured by her some time earlier. She has never seen one of these clusters made, for the reason that her mother died half a year before she was born. But within her is embodied an instinct for its fashioning and, at the proper time, it will assert itself and she and all her kind from Philadelphia to Buenos Aires will set about this very specialized task and will make structures very much alike of a proved architectural design, and well fitted to the purposes they are to serve.

When egg laying time comes, nature gives the mantis materials with which to work.   These are secreted from its

body and are not unlike those from which the cook makes frosting for her cake.   The mantis likewise has an egg-beater and works it most effectively. She develops a goodly quantity of froth. While this is still in the plastic state, she begins shaping her egg case. The part of the froth that is toward the bottom is heavier than that at the top.   She skims the light material off and puts it aside for a special purpose.   She makes the case, looking like a head of wheat, out of the heavier material.   She fashions the design very exactly, yet with seeming carelessness, since she never even looks at her work.   Then she places her eggs, some scores of them, like a braid of flaxen hair, down the middle.   She covers them with the lighter material that she has kept in reserve.   Her frosting, in contact with the air, gradually hardens.   Soon it is as tough and rugged as the shell of an almond.   The eggs have been put away for the winter.

EGG CASE OF THE MANTIS.

Nothing much happens until one bright day in the following June.   Then of a sudden this egg cluster begins to come to life.   Along the middle, where the lighter material was piled, a tiny creature with shockingly large eyes begins to show itself. Almost immediately scores of others appear. They work in squads. They are breaking through where the mother purposely made her structure weak.

There are perils, of course, for these tiny creatures. There is even a special chalcis fly that can penetrate the tough crust of the mantis' nest and deposit its egg within the egg of this sturdy builder. It devours the mantis

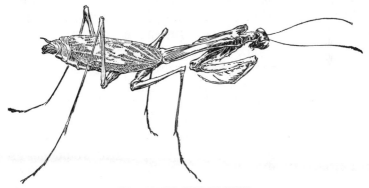

THE MANTIS LIES IN WAIT.

grub. So it happens that some of the little ones that emerge are not mantes, but chalcis flies.

And when the baby mantes begin to emerge there are sometimes hosts of tiny ants standing in wait. They fall upon the new-hatched and helpless little fellows and devour them in great numbers or carry them off to their burrows. Many more are eaten than escape. Those that do get away pass through one or two moults, develop a hardened crust, get their elbow pincers to work, begin on lesser insects like gnats and mosquitoes, and are soon lusty and aggressive. It is then that they turn the tables on the ants and the eaters become the eaten. It would seem that it might be that the massacre of their brothers and sisters, when they first appeared, had embittered the mantis, and that, through the rest of its life, it devoted itself to taking vengeance on the insect world.

Thus the testimony is produced that sets forth the life and character of the praying mantis. It would seem that the conclusion is bound to be that the creature is a hypocrite, is not the mild and benevolent sort of person that it seems to be. Instead of praying it is, in reality, laying in wait for whatever it may destroy. It is blood-thirsty, greedy, murderous. It is a wolf in sheep's clothing, an insect ogre.

In the face of this evidence, one would be inclined to feel indignation and enmity toward the mantis. As compared with the friendly grasshopper, for instance, there would seem to be no choice. Who would not assert his friendship for the genial fiddler of the meadows?

Yet, when the ends served by these two insect cousins are considered with relation to their bearing on the welfare of man, the conclusion is quite the contrary. The grasshopper is man's enemy, since it holds over him the constant menace of becoming so numerous that it will eat him out of house and home. The mantis, on the contrary, is man's friend. It eats nothing that is useful to him. No matter how greatly the mantes multiply, they would be no menace to man. In fact, they are most helpful to him. They eat his enemies. They keep down the number of those insects that would destroy him.

They are important in maintaining that balance in nature which makes of the world a pleasant place in which to live.   They fight beside man in the unending contest which keeps in check the insect hordes that might otherwise overrun the world.   They are stern warriors and some of their methods may be quite harrowing, but the results that come from their having lived are in the interest of the general welfare.

QUESTIONS

1. Can you suggest a reason for the peculiar beliefs of certain people about the praying mantis?   Tell any stories you have heard about strange beliefs concerning insects, birds, or animals.
2. Tell the class how the mantis is particularly adapted in each of the following ways to the life he leads:
    (a) Protective coloration.
    (b) Development of front legs.
    (c) Hinged body.
    (d) Ability to turn its head.
    (e) Protection by liquid.
3. Tell the story of the mantes eggs.   How does this story illustrate what is known as instinct?   Mention another remarkable illustration of instinct in the insect world.
4. What is meant by the "balance of nature"?   What illustration of the balance of nature do you find in the story of the praying mantis?
5. Classify the mantis as a friend or an enemy of man, and give reasons for your answer.
6. Make an outline of the chapter about the praying mantis.

Chapter XI

THE ANT

YOU may have an ant hill in your back yard, or a colony of carpenter ants in your woodpile. It is likely that they have been there for years and you have spent many an hour watching them at different tasks. But yesterday you saw a single file of them, like so many soldiers, marching away, empty handed, on some special mission, and another file, with plunder on its shoulders, coming back to camp.

You know these insects well, almost as well as the family cat asleep there on the door mat. You know, for instance, that they are plodders on six stout legs that never dream of taking to the air on wings. They no more have wings than has the cat. You cannot be deceived about that.

Yet one day in June you come into the garden and it is aswarm with flying ants. You think they must be strange visitors until you discover that they are coming out of the old familiar nest that you have watched so long. Your old friends, the wingless ants, are still going methodically about their business, but here is a great flurry of these ants with wings. What, you may well ask, is the explanation of this strange thing that has happened?

And the answer is that a peculiar rite, the master

122

celebration of the ant world, is this day taking place. It is a rite that can be witnessed but once in a year and you have happened to miss it in your garden heretofore.

These creatures that you see decked out in gauzy wings are not the industrious ants that you know so well. Your old friends will remain, will harvest their crops, milk their cows, will wage their warfares just as of old. These are the young queens of your sturdy workers and their mates their consorts. They have just come of age and are breaking away from the home of their parents and striking out for themselves. It is their wedding day. There is an instinct implanted in each of these young queens to go out into the great world and found a new colony for herself, there to reign, multiply, and become the mother of uncounted other queens that will found other colonies. She is obeying that instinct.

Nature put wings on the thistledown so it could bear away a seed which might find a fertile bit of soil for growing other thistles. It caused the cockle-bur to stick to the tail of the cow that she might switch it off in a far corner of the pasture, there to take root. She gave wings to these queen ants and their mates, lent them for but a single day, that they might fly away to establish new ant homes. Thus do these species seek to strengthen their position in the world.

You have never seen these winged ants before because they have been kept housed below ground and have been carefully reared, fed, groomed for just this occasion. The workers have taken care of them. You will notice now that the wingless members of the colony have come out as though to bid them godspeed. You may note that, where a queen or consort hesitates, the workers nip them on the heels, urge them on their way. They may be queens, but they have had free board here long enough. These workers have their rights. Soon all the bridal parties have flown away.

These queens would seem to have chosen weaklings as husbands. At any rate, the consorts are much smaller than the queens. It is often true in the insect world that the male is of little importance to the community and is soon discarded. You will remember that the female mantis ate her husband. In the ant colony the consorts that appear on this one day in the year are the only males. The workers, even the warriors of the brood, are females. They never become mothers, however, as the queens lay all the eggs.

A single colony of ants puts many of these bridal couples into the air this June day. All the other colonies of the neighborhood do the same. These flights merge into each other. Sometimes it happens that the air is full of flying ants, that there are actual clouds of them. Sometimes they fall into streams or lakes and cover the surface of the water for miles.

When the journey of these bridal couples is over, both drop to the ground. The weakling consort fares badly out here in a rough world. He has been accustomed to being fed and is unable to hustle food for himself. He falls victim to other creatures that like food of a spicy ant flavor. In a few days it comes to pass that every queen is a widow, for there is not a consort anywhere to be found.

Few queens survive, for that matter. Maybe one of

all the flight that left your back yard succeeded in setting up a successful establishment. If ten per cent. of those that start out founded colonies, the world would soon be overrun with ants. There are enough difficulties in the world, enough enemies of the ant lying in wait, to keep the number of colonies down to about what they should be, to maintain that balance of nature that we find every-where.

The chance queen, who made a success of her venture on wings, establishes herself in a tiny burrow in the ground or in a log, according to her nature. The first thing she does is to tear off her wings, which she never expects to use again now that she has made her marriage flight. Now she lays a few eggs. These she looks after with in-finite care. They hatch into larvæ and later spin cocoons about themselves. The queen feeds them with supplies that she brings up from within herself. The larvæ can be seen to put their mouths to hers and the food may be observed to pass down their throats. The mother often licks them much as a cow licks her calf. They require much nursing.

Finally the cocoons open and young ants are born. These ants are workers. They set about tidying up the place, taking care of their little sisters, and finally fare forth for food. As soon as there are enough workers to do the chores, the queen establishes her court and reigns in state with many retainers to wait upon her. Her business from then on is solely the laying of eggs, an oc-cupation to which she devotes herself during a long life, for the individual ant may live to be ten or fifteen years old, a very unusual age for an insect. The success of her colony depends on the number of her children. If it waxes large and strong, she is a queen who can participate in many of the marvelous exploits to which her kind resort and because of which they have come to be regarded as the most gifted of insects.

The carpenter ants of the northern states, big black fellows, and the agricultural ants of the south, equally big but red in color, are two outstanding types. The carpenter ants dig chambers for themselves in such places as the timbers of a barn, while the agricultural ants work underground. They dig out many rooms on many levels. Their homes are like the catacombs of Rome where the Christian martyrs used to bury their dead. In them they have apartments of state for the queens where they lay their eggs. There are many attendants who bear these eggs away carefully to the hatcheries. There are nurseries where the young ones are given watchful attention by the workers. There are supply rooms for the food and ordinary barracks for the common people. Every individual has her tasks. Each performs these tasks with skill and competence, yet with no supervision whatever. All in all it is an orderly and well-organized household.

SECTION OF CARPENTER ANTS' NEST IN A FENCE POST.

Whoever wants to study ants may arrange captive colonies for the purpose. A thin layer of dirt between two panes of glass may become the home of a colony and there they may be observed in the course of their normal lives.

The industry of the ant has long been set forth as an example for man to follow. It furnishes an equally good example of personal and community cleanliness. The foreleg of the ant is supplied with bristles which make of it a very excellent brush. One may often see a busy worker in the field climb upon a pebble and go entirely over its body with this brush, removing from it the marks of toil. More interesting, however, are toilettes ants make after dinner when they are ready for bed. At such a time each ant carefully washes itself, or it may happen that they "swap work," and take turns cleaning each other. The tongue is the wash rag. Ants lick themselves to keep clean in very much the same way that cats and dogs do the same thing. They seem to take particular delight in washing each other. The individual being washed will stretch out its members and give every

ANT MAKING ITS TOILET.

evidence of pleasure while a fellow ant does the scrubbing.

This cleanliness applies likewise to the home. Never a particle of anything unclean is to be found about the ant community. Drop anything messy in an ant hill and the sanitary squad is immediately called out to cart it away. Ant homes underground have little ventilation and are without the cleansing influence of the sun. They might easily become litters of filth, musty and unsanitary. They would if infinite care were not exercised. But these housekeepers are immaculate. Every suggestion of soil is given prompt attention. These creatures seem to know by instinct those lessons of sanitation which many human communities have not yet learned. Or, perhaps, those that did not keep clean have long since ceased to exist through the attacks of parasites and fungi.

But there is more exciting work than housecleaning in the ant colonies. There are, for instance, slave raids. There are in America two great races of ants that are given to this sort of exploit. There are the Shining Slave Makers and the Sanguine Slave Makers. They are red ants. They raid the burrows of black ants.

Some morning one may observe a great commotion in the colony of the Sanguine Slave Makers. The open ground about the gates is full of excited individuals. The red Slave Makers are marshalling their forces. Yet, running about among them, are the black slaves, captured as babies in former raids. They are as excited as the warriors, are ministering to them, are apparently hurrying them into action.

It is an odd thing that this army is without a leader. It is marshalled and goes into action without command, a thing that is regarded as impossible among men. No individual stands out above the others. Yet its plans seem well made and generally understood. When the hosts are assembled, it sets out definitely toward a certain point, a hundred yards away, quite a distance from the ant standpoint. This is no mere sortie upon an unprotected outpost as are some raids. The attack is upon the stronghold of the blacks.

The blacks seem to know of the impending attack. Sentinels have been posted. Barricades are being thrown up. Gates are being barred. Black fighters are hurrying up from below. Man meets man with only nature's weapons and the battle is on.

These fighters may lock jaws like bulldogs in a grip that cannot be broken. A second black may seize the leg of a red so engaged and sever it from his body. Half a dozen individuals may become entangled in one gripping mass. Knights of one faction or the other may meet on the edge of the battle and duel to the death. The manner of it is like the fights of King Arthur's day when individual prowess won or lack of it resulted in death.

In the end one side overcomes the other. The reds are more likely to win. When they do, they plunge into the burrows of the blacks and carry away the plunder in which they are particularly interested. This plunder is the young of the colony, the larvæ and the pupæ that are on their way toward anthood. Each warrior swings a baby black over its shoulder and starts for home. The pupæ in their cocoons look much like grains of wheat, and many observers have thought these ants harvesters of grain, while they were, in reality, kidnappers.

These infant black ants are given the greatest care in the colonies of their captors. When they are grown they become workers, but have all the privileges of citizenship. They live in the colony under identically the same conditions as the native workers. They are loyal and show no tendency to mutiny. Perhaps it is as in human experience that loyalty is an almost ever-present attribute of the blacks.

The Sanguine Slave Makers do not seem to have degenerated because of slave keeping. Perhaps they have not been slave keepers long enough. They still work vigorously beside their captives at their masonry or harvesting. The Shining Slave Makers, probably having had slaves longer, have become so used to the ease that comes with too much service that they have degenerated. They still fight well, but are without even the ability to feed themselves while in the presence of food. Put a dozen of them in a bottle with a lump of sugar and they will starve. Introduce a single black slave and it will feed and sustain them all. Thus are the ill effects of slavery demonstrated in the insect world.

You may stand beside an oak tree some summer day and note that a column of ants is marching up its trunk. If you should wonder what is the object of their mission and should watch them until you had found out, you would have discovered another story no less alluring than that of their slave making.

These ants are going out to milk their cows. These cows are the plant-lice or aphids, creatures much smaller than ants, which drill holes in the bark of the tree and stick close to these wells, drinking sap all their lives. They are big cousins of the scale insect and little cousins of the cicada. They are in a way dependent upon the ants. Ants keep their enemies, the ladybug bettles, away. Often they take aphid eggs and their young into their burrows and care for them all winter. You may sometimes turn over a stone and find beneath it ants with these aphid young. They will grab the aphids and scurry away with them.

The aphid interests the ant in helping it by keeping on hand a supply of honey. The ant is very fond of honey,

AN ANT MILKING AN APHID.

as every housewife knows. The ant milks the aphid for honey just as man milks his cows. Watch one of these ants climbing the oak tree. Out on a limb by the base of a leaf where the bark is tender for drilling, the aphids will be found, drinking sap. The ant approaches the aphid and finds that, out of its abundance of food, it has secreted some honey to pay the ant for the care it has shown. The ant takes this drop and waits, and soon another drop appears. This is swallowed. The aphid now hesitates, thinking, possibly, that it has paid its bill. The ant becomes persuasive. It knows that most animals like to be stroked, so, with its antennæ, it begins to rub the back of the aphid. This gets results and other honey drops follow.

The ant visits many aphids and, as it does so, its abdomen becomes more and more distended. If you watch the ants that are coming down the oak tree, you will see that they are quite different creatures than those that went up. The difference is in the size of their stomachs. They have become, in fact, strutting honey bags.

Then, if you investigate carefully, you will make an additional interesting discovery. At the foot of the oak tree, underneath the leaves, there are established certain ant restaurants. There many workers are waiting for their dinner. Labor in the colony is divided. These aphid milkers are specialists. While they have been milking, other workers have been doing the housework, taking care of the babies, excavating for the new extension of the colony. These workers are not required to hustle food for themselves. They are fed in the restaurant. Many of them are fed by these honey milkmaids who no more eat all the honey they get than does the farmer's daughter drink all the result of her visit to the barnyard. These animated honey jugs go about, place their mouths to those of other workers, press the bulb of their stomach, and lo! The hungry ant is fed.

Probably the most remarkable group of all these honey gathering ants is one to be found in the Rocky Mountains.

It resorts to the very novel expedient of converting certain of its own members into honey jugs and hanging them up for the winter. These Rocky Mountain ants dig caves for themselves in the sandstone. One room in one of these caves might be six inches long, four inches wide and one inch high. This might be a honey storage room. Excavating it, the investigator would find that its roof was entirely covered by strange amber globules about the size of grapes. Careful examination would show that these globules had legs and heads, were, in fact, ants with greatly distended abdomens. These abdomens are well filled, as is found upon examination, with a very delicious honey. These living honey jars hang upon the ceiling of these caves through the winter, and the members of the colony, as rations grow short, visit them, get their sparing rations of honey, and so live through that season during which it is impossible for them to go outside and get food.

In Mexico these living honey jars are gathered, are considered a great delicacy for human consumption, and are served on such occasions as at wedding breakfasts.

Though ants are very largely lovers of honey and likely to live chiefly upon it as a food, there are certain varieties which resort to agriculture. There is in the southern part of the United States, for example, an agricultural ant of a very high degree of civilization. Its members go out in the summer and gather various grain and grass seeds and store them in their granaries below ground. There is one grain which is found to be actually cultivated about their nests. This is known as ant-rice, and the members of the colony very carefully remove all other vegetation from about their nests, thus giving it the best possible opportunity to grow and yield good returns.

All these grains are stored below ground, and it is again evidence of the ant's fondness for sweets that it allows them to become dampened and to sprout before

THE FAMILIAR ANT HILL.

it eats them. Under the influence of this dampness the starch in the grain turns to sugar. When these sprouts appear, however, they are nipped, which has the effect of maintaining this dampened and swollen condition in the rice. Thus does it become a palatable ant food.

A very common sight in some communities is that of

processions of ants marching past, evidently homeward bound, with fragments of leaves held in their nippers. Sometimes these fragments are so large as to cause the tiny harvester to become quite top heavy.

Observing such a procession one might logically conclude that these ants are carrying these leaves into their burrows to use as food. This conclusion, however, is not correct. These leaves are not eaten at all. These ants are gardeners of a quite different type from those that harvest grain. These leaves will be put down into beds underground, will be moistened, and will then be converted into muck plots in which these ants can grow a product of their own of which they are very fond. This product is mushrooms, for which they maintain supplies of seed and which they grow as skilfully as any human specialist.

The different varieties of ants are very numerous, and different groups resort to different practices, many of which are intensely interesting. Those here related are but typical. They are a very social group of insects and always live together in colonies which are likely to be very highly organized. Many of the acts of these groups are such as to cause the observer to marvel at the apparent intelligence shown and to raise a very delicate question as to whether these acts are purely instinctive or due to actual thought taken by these insects. Any ambitious amateur naturalist who wants to contribute to the sum total of human knowledge may find for himself an interesting field of experiment in attempting to demonstrate whether or not the ant is able to take independent thought.

Ants, generally, are a nuisance to man and should be put in the class of his enemies. They injure his meadows and lawns, weaken the timbers in his buildings, and spoil the food in his pantry. There is an incidental service which they render to him, a service which he does not at all appreciate, that may pay well for all this annoyance. They stir up the soil for him. They bring fresh earth to the surface from far underground. They make holes into which water may flow to soften up the soil. They work and stir up the ground, which makes it grow better crops. The earthworm is man's chief cultivation aid, but the ant is a not unimportant assistant.

But accepting the theory that the ant is a nuisance, as it undoubtedly is under many conditions, it can often be destroyed by pouring boiling water from the kitchen kettle into its burrow. More ambitious colonies can be killed by pouring carbon bisulphide into their holes and then closing them by tramping in dirt. The carbon bisulphide becomes a gas, goes all through the chambers, and suffocates the insects.

QUESTIONS

1. (a) Have you ever observed ants at work in your neighborhood? Why do they merit the praise bestowed upon them in the Biblical saying, "Go to the ant, thou sluggard; consider her ways and be wise."

   (b) Make a list of the civic virtues displayed by ants. What story might you tell little Johnny who forgets to wash his face and brush his hair?

2. (a) The European nations have important colonies in various patsr of the world. Many men, feeling an inward urge to go to less populated lands and start anew, migrate to strange places. Insects, too, respond to this inward urge, to this wanderlust spirit.

    (b) How are some ants equipped for this purpose?

    (c) Describe fully the preparation in the colony for this eventful trip.

    (d) Why does the "queen" occupy such an important place in the ant world?

    (e) How is the colony governed? Would you call the form of government imperialistic or socialistic?

3. Compare the carpenter ants of the northern states with the agricultural ants of the southern states. Consider their color, home, food-getting processes, care of young, division of labor.

4. (a) Describe the "slave raids" in the ant world. Why does the author liken the combats to those of King Arthur's day?

    (b) Which office in the ant world would be most desirable in the event of a raid? Why?

    (c) When human slavery was tolerated, many cruelties were inflicted upon the poor slaves. How do ants treat their slaves?

    (d) Do you approve of slavery in the ant world? Mention one harmful result. Is this true of life in general?

5. (a) Why have the aphids been called the ants little green cattle?

    (b) Tell of ants' persuasive methods of getting more honey.

    (c) Prove that this connecting bond between the ants and the aphids is of mutual benefit.

    (d) Why is this relationship hurtful to man's welfare?

6. The ants are classified among the enemies of man. Do you consider this a wise classification?

Chapter XII

THE HONEY BEE

HE old queen was in a towering rage, was storming about the hive, was quite upsetting the whole establishment.

The younger bees had never seen her like this before. They had known her only as she went quietly about the business of laying her thousands of eggs that had bred the many workers that had made the colony strong. It had been so peaceful there in her royal quarters throughout the season. The work of the harvest had proceeded without interruption. Stores of honey had been laid up, and the colony had grown to be one of the lustiest in the whole neighborhood.

And, young bees that they were, they knew the meaning of this royal rage. They knew it by an instinct that was planted within them. This queen that they had served so well, they were quite sure, had murder in her heart. No other passion burns with that peculiar fervor as does this desire to kill. And none is more bitter than a passion aroused by jealousy, and they knew that it was jealousy that stirred the old queen.

These young bees had heard the piping. An hour ago it had sounded throughout the hive. They had never heard it before, but they knew what it meant. A young queen was preparing to emerge from the cell where they had been so carefully providing her with that royal diet which caused her to grow to be a queen. It is only such queens at such times that do this piping. Bees are otherwise voiceless, but these emerging queens whistle through the slits in their sides or vibrate their wings, thus sounding the note peculiar to them.

The old queen had heard the piping or maybe it was the "queen odor" that warned her. She knew that a rival was about to appear in the colony. To be sure, this rival was her own daughter, but she was none the less consumed of jealousy. She would kill this interloper.

Worker bees, however, are not without authority in the hive. They had developed this new queen on their own responsibility. There are times into which new queens may be born when the old queen would be allowed to have her way, when she would be permitted to commit this foul deed which she contemplated. If the hive were not well-filled with bees, there would be no use to which a new queen might be put, and it would be as well that she should be killed. This hive, however, was overflowing with young bees. Perhaps it needed another queen.

In that belief the worker bees hurried to the apartment of the piping young queen. They threw their bodies in great numbers upon her door, made it impossible for the old queen to carry out her purpose.

It is the law of the tribe that two queens cannot live at the same time in the same hive. The old queen, therefore, calling her followers about her, organized a pilgrimage. They would go on a journey, establish a new home.

So it comes about that bees "swarm." Some thousands

of them come crowding out of the hive, cluster about its doorway. Then the old queen emerges, takes to her wings, and all these thousands group about her and the flight is on. The worker bees have brought this about by developing and protecting the new queen. They have done it deliberately because it was getting too crowded in the hive.

This swarm usually goes but a little distance and settles on some bush or the limb of a tree, pauses, as it were, to work out its plans. The beekeeper is likely to take advantage of this halt, to present a satisfactory hive, and to induce the swarm to become a part of the group that is supplying him with honey. If he does not do this, the swarm takes a second flight, usually to some such natural home as a hollow tree, and thereafter become wild bees.

HIVE OF THE HONEY BEE.

If the home is to be a hollow tree, these bees, after taking charge, give a demonstration of their ability as masons by filling in the doorway until it is the proper size. The hollow is probably reached through a knothole. This knothole may be too big, may expose the home too much to the weather. What they want is a narrow slit of a door through which but a few bees at a time can crawl.

So workmen are sent afield. You may have noticed that about buds on trees there is to be found a sort of sticky glue. The bees know this also. They gather this glue and use it as material for filling in their knothole doorway. Soon they have it just to their liking.

There is a well-established division of work in the colony. While this first squad was gathering the glue, the moppers-up have been putting matters to rights inside. A third squad has been given a much quieter assignment. Bees gorged to bursting with honey are selected that they may serve another and very peculiar mission. They are to become the wax factories of the hive. Comb must be provided before honey can be stored and comb is made of wax.

These gorged bees hang themselves up as a curtain, one clinging to the feet of the other, stomachs exposed. Nature, in her resourcefulness, causes the digestive organs of these bees to extract the wax from the food that they have been eating, and to secrete that wax through glands on the outside of the abdomen. This is done a good deal as the mantis secretes the material for making its nest when it is needed and as the caterpillar secretes the silk for spinning its cocoon.

The wax forms on the lower side of the abdomen of these bees and is scraped off by the wax workers who are to make the honey comb. It is hard and brittle wax, a good deal like chewing gum that has been stuck beneath a chair. The wax workers chew it as one might such a piece of gum. This softens it and gets it into a condition where it can be handily worked.

The wax-working bees start to building cells up in the top of their new home. Everybody has seen honey in the comb, but not everybody has taken the pains to notice the form of these cells. They are six-sided. Why, do you suppose, are they six-sided, instead of five-sided or seven-sided, or round? The answer is because six-sided cells are a better proposition from an engineering standpoint. Most of our lead pencils are six-sided for the same reason. Wood cuts up into six-sided pieces with no loss whatever. Take six six-sided lead pencils and you will find that you can so fit them together that they will merge into one bigger piece with a hollow through it. A seventh six-sided pencil will just fit into that hollow. They would have been cut out of the bigger piece with no loss of wood. There are no empty spaces between them. For round or five- or seven-sided pencils there would have been loss and they would not have fitted together.

So would it have been with the honey cells. The only other alternative would have been four-sided cells, and these, any engineer will tell you, would have lacked the strength of the six-sided ones.

One set of cells backs up against another. Their bases, however, do not so fit that one might open into the other. If they did, the whole would be much weaker. No engineer can design a plan for building wax cells that would be stronger or more economical of wax.

The hive is now ready for the honey gatherers. It is probably true that no creature on earth has a better worked out system of storing food for its young and for the winter than has the bee. The bees, obviously, are social insects. They live together in groups of considerable numbers. These groups are organized. There are queens, drones, workers. There are masons, harvesters, nurses.

This organization of colonies is very unusual. Most

insects live their lives as individuals. The ants, as we have seen, are social insects. Their colonies are organized. They store food, are fond of honey. It begins to appear that the ants and the bees are a good deal alike. Come to think of it, their forms are very similar. Both are wasp-waisted. That draws in another insect, the wasp. Certain wasps and certain bees are so much alike that you can hardly tell them apart. Most wasps also live in colonies.

Yes, these three insects are cousins. They form one of the most remarkable groups of all the insect world. They are the membrane-wings, the Hymenoptera. They are the man-like insects. Their manner of community life is so parallel with that lived by man as to be at times most startling. They are very admirable creatures. Cleanliness is a cardinal virtue with them. Industry seems to be the mainspring of their existence. Their efficiency appalls the student of their methods. They seem to have government—socialistic government where every individual works unceasingly for the welfare of all. Since literature began the ant and the bee have been set up as models for man to follow.

Thus far in organizing this new home the glue-gatherers, the moppers-up, the wax-producers and the wax-workers have been industriously at work. The colony has gone as far as making those six-sided cells that form the structure for the bee home.

These cells are intended to serve two purposes—that of vessels to hold the food supply of the colony and as nursery compartments for the young. The care of the young is a matter of administration within the home, but before it can go forward it is necessary that the harvesters be sent forth to gather food.

Those bees whose business it is to feed the colony get their honey from an entirely different source from that which supplies their cousins, the ants. Where that other insect, the aphid, bribed the ant with drops of honey to take care of it, the flowers step forward and offer similar benefits to the bee if it will visit them and perform one small service without which they could not exist. The flowers, as we have seen in our study of the bumblebee, must be fertilized—the pollen from one flower must be carried into the heart of another or that other cannot ripen into seed. The flower places a drop of honey in its depths as a bait to the bee which, having visited other flowers, is sure to have particles of pollen sticking to its wooly head. It is estimated that in the world there are one hundred thousand kinds of plants that would not exist but for the fact that the bee thus goes about carrying pollen from one to another.

This fee which the flower pays the bee is in nectar which, with a little evaporation of the water in it, is turned into honey. The bee gathers this liquid and puts it into a nectar sac which it carries about for the purpose. When it gets back home it will empty this jug of nectar into the honeycomb, there to be reworked into finished honey by the squad that is assigned to that particular task. Thus is the clear honey provided for the colony.

As other harvester bees, a younger group of specialists, go about, however, they gather yet another food product of great value to the hive. This is none other than stores of the same pollen which they carry about from flower to flower. The pollen is gummed up a bit with honey and is deposited in pollen baskets on each hind leg of the worker. When the boots begin to get unduly heavy, the harvester knows that it is time for a flight homeward, rises from the flower, and makes a "bee-line" for the hive. Students of bees have learned to find their hives by watching the course that these heavy laden bees take.

This pollen mixed with honey is stored in cells by itself and is a product quite different from the clear honey. It is known as bee bread and is a staple food in the rearing of the young.

As the workers in the newly established home thus settle themselves to their various tasks, the queen bee proceeds to perform her outstanding mission, that of laying eggs from which young bees are to be hatched, thus to strengthen and develop the colony. Certain portions in the hive, sometimes half of it, are set aside for nursery purposes. The scheme of architecture is the same here as in the storage rooms and consists of these regularly arranged six-sided cells. In one cell after another the queen deposits an egg. She seems to consider that her responsibility ceases when she has performed this task, and leaves the egg entirely to the care of another group whose specialty is to look after the nursery.

As in the case of the ants, so with the bee is a great deal of care given to the young. Soon this egg hatches into a larva and this tiny grub is watched over by its nurse and is supplied with bee bread at intervals as frequent as it will take it. The manner in which the nurse administers this bread is a good deal like that of the mother bird who brings the worm to the nest of her young where she finds hungry and wide-open mouths into which she drops it. The effectiveness of the care is shown by the fact that in five and a half days the larva is 1500 times the weight of the egg from which it came.

This care of the young is another respect in which this order of insects, the ants, the bees and the wasps, differ from most others. The butterfly, you will remember, pays no attention whatever to her baby caterpillar. The mantis dies long before its young is born. The bee, manlike again, cares for its offspring until it arrives at maturity. Like most other insects, it passes through the larva stage and the pupa stage, during which latter time of sleep it is sealed up in its cell by its nurse and left in quietude until it has transformed into a bee, at which time it eats its way through the end of the cell and emerges.

The great majority of these young bees are workers. They are, in reality, females, but are not fully developed. They do not mate and become mothers. The queen is the only mother of all the thousands in the hive.

In addition to the workers, there are other inmates of the hive known as drones, because they do no work or because of the droning noise they make. These drones are males, the only males in the hive.

The queen can produce drones at will. The eggs that make workers of new queens are fertilized, but when the queen decides to produce drones, she omits fertilizing the eggs. These unfertilized eggs are provided with special compartments much bigger than those used by the young worker bees, as the drone is the giant of the hive. He emerges in due time and thereafter idles about the place, gets in the way, suns himself on the front stoop from twelve to three on bright days.

One of the most remarkable miracles of nature is the ability of the workers in the hive to produce a queen bee

THE BEE MAN AT WORK.

at will from the very same eggs that ordinarily yield only commonplace drudges. These workers who, in the final analysis, are the rulers of the hive, begin to find that there are too many of their kind about. The place is getting crowded. And, besides, there is the instinct for colony planting which is present among all first rate peoples. Possibly it is time to make provision for sending out a mission, a swarm.

If there is to be a swarm, there must be an additional queen. Obviously, say the workers, it would be wise to develop a queen against the possible necessity of the approaching end of the season. So they take this ordinary worker egg, place it in a special compartment provided only for queens, minister to it, feed it (and here is the nub of the whole situation) upon a special food brewed by these queen makers. Under this care and feeding the worker egg grows to be a queen.

Then comes the day of the piping and the flight of the old queen and her following.

Thus is the cycle of the life of the bee pretty well covered except for the introduction of one bit of romance and another of tragedy.

The romance has to do with the young queen left behind after the swarming. Instinctively she realizes that hers is the responsibility for mothering this depleted hive, that it will weaken and die if she does not provide new generations of bees to carry on its work. Yet to be a mother she must find a mate, and here in the hives there are only her brothers with whom she might wed.

But all the mothers ahead of her have known what to do in the circumstances and instinct again urges her on.

Then comes a sun-bathed summer noon, glorified by a cloudless sky and undisturbed by rowdy winds. The young queen strolls to the hive doorway, makes her observations, decides that her hour has struck. It is on such days as this that princes and queens ride forth with Cupid, and she will this once stretch her wings and find what the fates have arranged for her. She rises into the summer air and takes her course up, up, up, as far as eye can follow her, up into the crystal blue, away from the pother of earthly things, into a solitude little likely to be disturbed.

And then and there occurs another of those marvels of the insect world. No sooner has the young queen taken flight than the fact is known in all the colonies of bees roundabout. Those lazy drones that have thought only of bee-bread and sunshine spring suddenly into animation. They hurry into the open as so many fighting planes might rush from their airdromes on the battle front when the purr of an enemy machine is heard aloft. Nobody knows just how the alarm is sounded, but the theory is that the escaping queen in challenge sprinkles perfume as she goes and that the marvelously sensitive antennæ of these insects pick up the message. Soon there are a thousand lovers entered in the lists for the favor of this one maid.

No one knows how the queen makes her choice. Possibly her flight is so high as to test the vigor of the stoutest of these pursuers and the reward goes to him who ascends to greatest altitude. Possibly her choice is left to a chance meeting and he wins who is luckiest in the course he has taken. Whatever the method of it, the queen mates up there in the azure vault, lives her hour with her lover and returns to the hive, not again to see him and not to leave it until she is displaced by a younger queen and goes forth to found a new colony.

After their flights in pursuit of the queens the other drones return to the hives, to their idleness. Only this one drone in a thousand who found his queen on that day of flight is ever to have a mate. Those that remain are useless. They get in the way of the workers, litter up the place. They are the only bees that are untidy. It takes four workers to provide for and clean up after each drone. Despite this, the workers take care of them with no show of resentment until the time comes when the honey-gathering season is over, the time from which the hive must begin to live on its reserve stores, must begin considering the supply that must be conserved to last through the winter. It is then that the workers again show themselves to be masters within the hive. They set upon the drones and drive them from the colonies. Some of them they kill outright. The others they turn into a world with which they are unable to cope. All perish. The curtain is rung down on these idlers who no longer have any possibility of serving a useful purpose.

The honey bee is a domesticated animal, just as truly as is the cow or the chicken. It has lived in man's back yards for thousands of years, has contributed bountifully to the supply of his table. It was domesticated in the far east long before the coming of Christ. America, however, knew it not until Europeans settled over here. The Indians marveled at the "white man's flies" that the colonists brought with them. Yet now in this country alone every year 25,000 tons of honey are taken from these hives.

At that, the honey is the lesser service the bee renders to man. In carrying pollen from flower to flower, though he knows it not, she heaps products in his lap worth a thousand times as much.

Were it not for the fact that bees carry pollen, there would be no apples in the barrel against the coming of winter. But for this fact, there would be no orchards that ripen peaches, pears, and cherries. In fact, there would be none of that fruit of the orchards that grows upon the leaf-dropping trees that we know so well. Neither would there be clover fields in which the patient milch cow ruminates so pensively. Neither would there be buckwheat for the breakfast hot cakes. Neither would there be scores, thousands, of the plants that are familiar to and helpful to man in his ordinary walks of life. The honey of a single great State is worth, to its producers, one million dollars a year: the apples in that same State are valued at fifteen times as much. So does it become obvious that the buzzing bee serves man more effectively in an indirect way than in merely placing honey upon his table.

QUESTIONS

1. (a) We generally think of queens as having much power. Is this true in the bee world?
   (b) What rights do workers enjoy?
   (c) What privileges do they exercise on the day of the piping when the new queen appears?
   (d) What treatment do the "workers" accord the "drones" at the end of the honey-gathering season?

2. (a) With whom are your sympathies, the new queen or the old queen who is obliged to make a pilgrimage?

   (b) How is the old hive benefited by this moving out? Compare this act with human conditions in old-world countries.

   (c) Why does the wise beekeeper try to attract the swarm of bees to his apiary?

3. (a) Illustrate the expression, "as busy as a bee."

   (b) Describe the work of preparing the new bee hive, the making of the comb, the manufacturing of wax, the engineering skill displayed by bees.

4. (a) Queen Apis is a homebody. She leaves the hive but twice in her life. Describe her flight into the high heavens.

   (b) The bees, not unlike the Brahmins, have solved the labor problem by employing the caste system. Make a list of the duties and privileges of the various castes, such as queens, kings (drones), workers, soldiers.

   (c) Describe in detail the work of the masons, harvesters, and nurses.

   (d) What is the significance of the statement, "He made a bee line for it."

5. Compare bees with ants. Which are more efficient, which more interesting, which more helpful to man?

6. (a) The bee serves man in two very important fields, illustrate.

   (b) Make a list of all the products which this highly domesticated animal gives to man. Estimate in dollars and cents the value of the bee to the farmers, truck-growers, dairymen, and orchard-growers of your state.

CHAPTER XIII

THE HUNTING WASPS

THE largest and most villainous-looking of American spiders is the fur-covered tarantula of the southwest, as big as a baby's hand, a creature that is wont to come out of its silk-lined well and stalk about the deserts as the master surveys his own domain.

But there are occasions when the arrogant mien of the tarantula changes, when it seems thrown into a great nervousness, when it cowers as a frightened creature, hunting shelter beneath any chip or shrub that it can find.

The change is brought about by the hum of a pair of wings. Watch and you may see their owner come into action—a black and yellow, trim, high-strung creature, giving evidence of physical energy in her every movement. She is a western species of the digger wasp, called out there the "tarantula-killer."

She circles above this huge spider which now runs in panic, seeking shelter, now rears itself to give fight. Huntsmen prize big game, game which fights and in the killing of which there is danger. There is more sport in killing a lion or an elephant than in killing a rabbit or a deer. These wasps seem to have the true sporting instinct, for they will go far out of their way for a chance to fight to the death with a tarantula six times their size.

The wasp circles with jerky, disconcerting movements of lightning quickness above her victim. The tarantula stands on its hind legs and strikes at its pursuer. So rapid are the movements of the wasp, however, that the great spider seems to become confused. This is the moment of opportunity for the wasp. She strikes. She drives her sting into the tarantula's body. In doing so she produces a very remarkable result. She paralyzes but does not kill her victim. As a matter of fact she injects a poison. The tarantula will lie motionless for weeks, but will not die. It is the wasp's method of keeping her meat fresh.

Digger wasps are rather peculiar members of their family. They are solitary wasps. They play the game of life all by themselves. You have doubtless seen one species many times darting about your lawn or about the roadside. It is very likely to have attracted your attention when it started to dig a hole for its nest in some dry embankment. It digs much as would a dog, but works more rapidly and excitedly, and brings into play its hind legs, which fling away the particles of dirt that its front legs loosen.

This digger wasp sinks its hole some six inches deep; then from it extends three or four side tunnels for a lesser distance, and makes a sort of circular bulb, one inch across, at the end of each tunnel.

It is into this tunnel and into this bulb that the wasp will finally succeed in dragging her big game. There she will lay an egg in its body. When this egg hatches out, the stored food will still be alive and, therefore, fresh meat. The grub will begin eating it, and, by the time it has been devoured, will be fat and full sized, ready to spin a cocoon and go to sleep for the winter.

Another big game prize for another specimen of digger wasp is the cicada, noisy harvest fly of the East. The wasp flits about the trees and finds this trumpeter, four times its size, sitting on the limb of a tree sounding its call. The wasp pounces on her sturdy victim and the two tumble to the ground in a wild flutter while her sting is being inserted. With the big cicada paralyzed, the problem becomes one of transportation. How is she to get it to her burrow?

Watch and you will see her laboriously climb up a nearby tree, dragging her game behind her. When she has gained sufficient height, she makes a flight, but finds that her wings are not strong enough to carry this great weight. She keeps up as best she can, however, volplaning gradually to the ground and, by the time she reaches it, she may have traversed half the distance to her nest. She climbs another tree, makes a second flight, and has accomplished her purpose.

A kinsman of the digger wasp and likewise a great huntress is the mason wasp, or dirt-dauber, particularly abundant in the southern part of the United States. To be happy and prosper dirt-daubers must have supplies of mud, must have access to buildings in which to make their nests, and must live where the spider crop is good. One sees them buzzing about some such place as a leaking water-trough where the tramping live stock has worked up a good supply of mud. They dig themselves out sturdy balls of this and fly away with them to a nearby barn. There, in a snug corner, they begin the erection for themselves of adobe houses such as the Mexicans build. They must build under cover, as these dirt houses would melt if they got wet. Each chamber in these houses is as large and as round as a .32 cartridge. There may be half a dozen cartridges.

This preparation made, the hunting begins. This dirt-dauber is also possessed of the true sporting instinct. She likes a good fight for her game. This spider-hunting offers enough sport to make it exciting. The spider is himself a huntsman or, perhaps, it would be more accurate to style him a trapper. He weaves a web in which his victim becomes entangled; then he captures it. The dirt-dauber delights in finding this spider in the act of taking game and in turning the tables on him by picking him out of his own dangerous web.

The dirt-dauber stings her victim, paralyzing it. She lays an egg in the first one she puts in each of her cartridge-shell chambers. Then she puts in other spiders until the chamber is filled. The food she has thus stored keeps fresh because it is still alive. She then seals up the cartridge, which pretty well converts the contents into canned spider. The egg hatches, the grub begins to eat, keeps at it until the spider meat is all gone, then turns to a pupa and sleeps until it is ready to come out a wasp.

These solitary wasps are very trim, clean, active creatures, but they are more primitive than most of their relations. The wasps, you will remember, are cousins of the ants and bees, which are the smartest and most highly developed of all the insects. They are social insects, live in communities, have government and all that sort of thing.

There are also social wasps with community life. There are, for instance, the hornets and the yellow jackets. They live in colonies, practice a division of labor, give personal care to their young, and otherwise conduct their affairs much as do the ants and bees. They are more highly developed than the solitary wasps.

The similarities and the differences among these related insects are most interesting. There is no question that the ants, the bees, and the wasps all came from the same grandfather if they could be followed back far enough. A man, before the Revolutionary War, might have had three sons. One became a Gloucester fisherman, and his descendants may have ever since been fishermen. One may have pioneered into the Tennessee mountains and his descendants may have ever since been mountaineers. One man may have become a University professor and his descendants may have lived ever since in an atmosphere in which study ruled. These three families would have grown to be quite different sorts of people. They would be different in appearance, would ·live in different sorts of houses, eat different food.

So did these insects come of a common ancestor, but, living under different conditions, they gradually grew to be different. They have been growing apart for hundreds of thousands of years and there has been a generation every year. Yet the differences between ants and bees have probably come about much more slowly than the differences between the Tennessee mountaineer and the Gloucester fisherman. The ant, living mostly underground, found that it had no use for its wings and discarded them. The bee, living among the flowers, got the habit of drinking honey and ceased to be a huntress as it once was. The solitary wasp has changed least, still roams the open spaces, and devours its kill. The social wasps, living in communities, have acquired a liking for honey, may be seen sipping fruit juices in the orchards, but still feed largely on insects. They are making progress toward becoming honey-eating creatures. The bee went entirely to a honey diet when it decided that it would try to live through the winter. It had to have food to do this; so it began storing up honey. The social wasps, with the exception of the queen, are content to die when the cold weather comes in the autumn. They eat honey and feed it to their young, but do not store it. They have traveled half the journey from being flesh-eaters to being honey-eaters.

The ants, becoming agriculturists rather than huntsmen, many of them have lost their power to sting. The bee stings only in self-defense and has become so ineffective at it that it cannot extract its weapon and is torn and killed. There was a time, undoubtedly, when the bee used its sting as a lance in slaying the creatures it used as food just as the wasp does today—a lance which it could thrust home, withdraw, and use again.

These social wasps deserve renown among the artisans of the world. They are the inventors of one of the great commodities extensively used by man, yet unknown to him until a few decades ago. They made the first wood-pulp paper, news print, that material upon which your daily paper is printed. They have been making wood-pulp paper for tens of thousands of years; while man used rags in paper manufacture until a few decades ago. They have been building their nests of this paper. The modern manufacturer takes logs from the woods, grinds them up, converts them into pulp, which is a mass of wood fiber, puts in the necessary binder, or filler, which may be nothing more than clay, and rolls the pulp out into sheets of paper. The wood fiber holds the other materials in the paper together as straw holds the dirt together in the making of bricks. The wasp goes to the fence, pulls fiber after fiber off of it, chews that fiber into a pulp, mixes it with saliva which contains a binder, and rolls it out into sheets that form the walls of its house. Man could have learned this method of paper-making from the wasps centuries ago if he had taken the trouble.

Everybody has seen wasps' nests. Those that most people have seen are the nests of the lesser wasps that build in the barn or under the eaves of some outbuilding. The wasp starts by making of this same paper a stem by which the nest is to hang. This it attaches beneath some timber. The nest begins to grow upon that stem. First there are three cells. They are six-sided cells like those in the comb of the bee, as the engineering advantage of building them that way is a matter of knowledge to the whole family of the membrane wings. More and more cells are added until the nest becomes, by autumn, eight inches across and contains hundreds of cells.

Yet this is the simplest of the nests of the social wasps. There is the king of the wasps, the bald-faced hornet, which attaches its nest to the limb of a tree. The bald-faced hornet is a paper maker. He starts his nest just as does the common wasp. But when it has grown half a foot across, the hornet exends the stem and begins another layer of cells. There may be half a dozen of these layers, one below the other. The hornet also starts an umbrella over its nest, as do the tree wasps and the ground wasps. This is brought down at the sides until it becomes a paper bag that encloses the whole nest. There is a hole in the bottom through which the wasps gain entrance. There are two or three layers of paper in these walls with air chambers between. Walls with air chambers between constitute the most scientific method of maintaining an even temperature within them. That is the principle upon which the thermos bottle is built. These wasps are scientific also in the application of methods of ventilation. In modern theatres great fans are placed at proper points to send currents of air in or out. The wasps and the bees use the same principle. When it gets stuffy in the nest, the ventilation squad is hurried to the doors and start their wings working to drive a current of air through the establishment.

The ground wasps develop very big colonies. The only outward evidence of their existence is a hole about an inch across coming to the surface. Watch the outlet of one of these big nests at the rush hour and for ten minutes count the wasps coming out and going in, and you are likely to find that three or four hundred have passed each way.

It would be a shame to kill a colony of these wasps to gratify the curiosity of one individual, but maybe it would be worth it for a class. This wasp colony is probably worth as much to the farmer as old Bossy yonder who gives milk for his family. In the season it destroys vast numbers of insects that would otherwise feed on his crops. If the screen is left open its members will even come into the house and kill the flies that annoy the family. Wasps are regarded as enemies to man because, when attacked or frightened, they use their stings. As a matter of fact, they are man's friends and helpers. They work diligently in keeping the balance in his favor by killing the insects that are his enemies. Aside from an occasional sting, they do nothing that is harmful to him.

But if you should think it wise to sacrifice the ground wasp colony in the interest of research, here is the way to go about it. Go to the nest by night while nearly everybody is at home and pour a few spoonfuls of carbon bisulphide into the hole. Tamp a handful of damp clay over the mouth to keep the vapors in. The liquid will turn to gas and suffocate the sleepers. Return early in the morning before those few who stayed out all night are astir, dig a trench two or three feet deep beside the nest, then carefully chip into the earth where it is located.

Presently you will open up a hole as big as a half-bushel basket. At the top of it there will be a root and the stem of the nest will be fastened to this. That is where the queen wasp started early in the spring. It was a very modest little cave, as big as a teacup at that time. The queen wasp did all the work, took care of the babies, foraged for food until the first workers grew up. Then they took hold of the chores, and things went faster. The practices of the tribe were followed in supplying nurses for the young in giving them predigested food from the nurse's mouth. In the case of the social wasp, half civilized from the bee standpoint, this food was sometimes honey and sometimes minced insect which the nurse had eaten some time previously.

These workers found that they would need more space underground that the home might be enlarged. They became miners. Little by little they dug out the dirt and carried it entirely away. There is no evidence of it about the door of the home. They kept the hole amply large for the growing family. Below the nest there is always a considerable space. This is used as a dump, for the ant and bee idea of cleanliness prevails with the social wasps, and there must be some place for the waste.

It is no great task to capture the nest of a wasp or a part of one and keep it in a glass case for observation. I will leave methods to your own ingenuity. The captives become honey eaters to the exclusion of all other food and thrive on it. You can watch these brusque warriors at the kindly task of ministering to their young ones. Day by day you will observe the growth of the grub. The wasps will scrape wood fiber from a block you have put in the case for that purpose, will chew it into pulp, make paper of it, and enlarge the cell for the growing grub. Finally, when it is ready to enter the pupa state, to go into the long sleep from which it will emerge a wasp, a neat paper cap is built over the top of the cell.

But an odd thing happens at the end of the season. It is in the nature of wasps that they shall not live through the winter. Though the temperature in your cage is not low, though there is still plenty of honey in the dish, the spirit of life in your colony begins to let down. The nurses begin to neglect the grubs. They even come to their own meals with halting gait. Their clothes begin to look dusty. They no longer brush them with their accustomed care. Finally the lagging nurses turn upon the charges. They go into the cells, seize the grubs, tear them from their beds, drag them out, fling them ruthlessly to the bottom of the pit. It is a massacre. The nurse wasps seem to realize that it is too late to bring these children into the world and to forestall an impossible situation by killing them. But a few old hag wasps remain. These, in turn, feel the approach of the end. The instinct of good housekeeping is still upon them. Each in turn comes out of the nest and topples in the final plunge into the pit. In death they will not defile this apartment that they have kept so spotless during life.

Finally the last wasp is dead. If the nest were outside, however, there would be a few individuals to escape, the young queens. Some weeks earlier they would have taken their flights into the blue, there to meet their mates and then instinctively crawl into warm nooks there to nurse the spark of life which, in the spring, would warm into flame, into that pulsing, foraging, constructive vigor that would make competent provision for other generations of wasps.

QUESTIONS

1. (a) The hunting wasps, unlike their relatives, the bees, are not vegetarians. What kind of diet do they provide for their young?
   (b) In what way is the digger wasp a good sport? Compare her size with that of her prey, the tarantula.
   (c) Why does she lay her egg in the paralyzed victim's body?
   (d) As the blood-thirsty spider shows little mercy to its victims, we might regard the wasp's behavior as a form of retributive justice. However, this is an unsportsmanlike way of killing. Explain.

2. (a) The digger wasps enjoy hunting big game. Describe the combat between the wasp and Cicada.

   (b) How is the paralyzed Cicada carried to the wasp's nest? What difficulties are surmounted on the way?

3. (a) The mason wasps, or dirt daubers, knew how to build long before Solomon constructed his temple. How does mother mason construct her abode-like home?

   (b) What provisions are made to house a large family?

   (c) The mason wasps toil unselfishly from morn till night for the housing and care of their offspring. How is "canned spider" prepared for the grub.

4. (a) Compare the social wasps (hornets and yellow jackets) with their cousins, the ants and the bees.

   (b) Explain the theory of the common ancestor of ants, wasps, and bees. How has each adapted itself to its environment? Note the interesting changes that have been brought about as a result of adaptation to environment.

5. (a) Man can learn much from Mother Nature. Describe the highly developed skill of the social wasps along manufacturing lines.

   (b) Compare their work with that of man.

6. (a) Wasps' nests offer an interesting study. Compare the making of the common wasp's nest with that of the hornet's nest. How are the ventilation and heating problems solved?

7. In what way do wasps help in maintaining the balance of nature?

CHAPTER XIV

THE SILK WORM

THERE are not many products with which one comes in contact that have back of them five thousand years of continuous development as has the silk of commerce.

Automobiles, typewriters, railroad trains, electric lights, radiators, mowing machines, sewing machines, steamboats, are all baby inventions that have come into existence in the last hundred years. A thousand years ago printing, sugar, glass, had never been conceived. The Romans knew no foot covering more ambitious than a sandal. Two thousand years earlier and man was without any metal which he might fashion into a tool. Europe had witnessed no glimmer of the light of civilization. Yet one may go even beyond that and, in the Far East, silk was being woven into cloth.

And the basis of this ancient industry is a baby insect that is not even allowed to grow up, that is killed before it is out of its swaddling clothes. It is the silk worm, which is not really a worm at all, but the caterpillar of a delicate little ashy-white moth. It is a caterpillar that never becomes a moth, but dies that silk may be provided for those dominant man creatures of the world. It stands in a class by itself as being an insect which is directly responsible for one of the great industries of the world.

165

The silk industry was born in China before Rome, Athens, Damascus, even Egypt, came into being. Over there the legend of its birth is well preserved. It happened in the time of Emperor Huang Ti who reigned in that hundred years of the twenty-sixth century before Christ, and who was himself by way of having the name of being a great man since he is said to have first conceived the calendar. But greater still was the fame gained by Si Ling Chi, his Empress, because it was she who gave the silk industry to China.

Si Ling Chi began as a child-wife, for, in the Orient, they take their brides young and let them grow up afterward. As a child she played in the emperor's garden and there observed the caterpillars that came out in the spring and lived on the mulberry leaves. Imagine her delight when these caterpillars began unwinding yarn from their mouths, kept at it unceasingly for three days and nights, and in the end had quite enveloped themselves in it. Imagine this little girl playing with the cocoons that resulted. Imagine her then dipping one of them into the hot tea that is sure to have been served to her in the garden. Imagine this tea softening the gum of the cocoon just as warm water does today all over the world where cocoons are unwound. Imagine the little Empress taking up the end of this silken thread and beginning winding it around her hand. Imagine the cocoon turning over and over in the cup of tea like an unwinding spool of thread as it surely would have done. Imagine the little almond-eyed Empress of the long ago going on and on to see how long the thread would be. She probably never came to the end of it, as some of these threads of silk are a half mile long. But she probably unwound enough of it to marvel at the fineness of the texture of it and to see a feminine vision of the cloth into which it might be spun.

At any rate, the Chinese say that, when this child-wife of an Emperor grew up, she invented a reel on which to wind the silk from cocoons, gathered quantities of it, actually made it into cloth, encouraged her subjects to do likewise, and thus started an industry on the way toward a development that would eventually gird the globe.

For three thousand years, a period longer than the life of our Western civilization, China kept the secret of silk culture to herself. It was some three hundred years after Christ that certain Koreans stole four Chinese girls from their native land and spirited them away to Japan where they taught the mysteries to a neighbor nation. So prized by the Japanese was this feat that today there stands in the Province of Settsu a monument to these four Chinese girls. It was 550 A.D. that Justinian, great Byzantine Emperor, commissioned two Persian monks who had dwelt long in China to return to the Far East and bring back some of the seed of the silk worm. This they did, hiding the eggs in the hollows of bamboo canes they carried, and in Constantinople hatched them out into silk worms which grew and prospered and became the parents of all the silk worms that South Europe knew for fifteen hundred years, and the source of the silk that has clad the rich and the powerful of that region for all the occasions of splendor that have since intervened.

The Orient and South Europe are still the sources of

the silk of the world largely because they have an abundance of cheap labor, the unwinding of cocoons being a matter of infinite detail.

But to get back to the insect itself. The silk worm is the caterpillar of a moth. You will remember that the moths and the butterflies are cousins, that they belong to a family known as scale-wings. The butterflies go out by day and the moths by night. The caterpillar of the gipsy moth has of late been busily eating up all the trees of New England, is a frightful enemy of man, as, in fact, are most moths, the clothes eating moth, for instance.

Moths are very abundant and widely distributed in the United States and nearly everybody has seen many varieties of them, some of them spinners of cocoon silk. One of the handsomest of them all is the Luna moth, a big pale-green, swallow-tailed creature that develops from a very fat and ugly caterpillar, usually found in oak, hickory, and chestnut woods. Another well known species is the Cecropia moth, which is very large and splendid, sometimes measuring six inches across the wings. It is reddish brown and as shaggy as a Shetland pony. Toward the point of each wing it has that same sort of "eye" which one sees on the peacock's feather. There is a very peculiar thing about this "eye." It appears on many creatures throughout the animal kingdom. The leopard, for instance, has it on his fur. It is seen on certain flowers. It is even present upon certain fishes. It is so widely distributed that one might almost take it to be a sort of trademark of Mother Nature.

Quite different from these gaudy moths is the shabby yellowish-brown creature that is the mother of the Army

Worm which sometimes creates great havoc by devouring grain and grass crops.

All moths come from caterpillars, which are odd and interesting creatures, and which most people have an abundant opportunity to observe. They are customarily referred to as worms, as in the case of the silk worm, but, as a matter of fact, they are not worms at all. Correctly speaking, the worm is a soft, jointed creature which remains that sort of creature throughout its life. The earth worm, for instance, is properly named. The silk worm, however, or the army worm, are moths in the larva stage and soon change from worm-like forms. So is the measuring worm, which, by the way, does its "measuring" because of a peculiar habit into which its ancestors fell. They had feet on most of the joints of their bodies like other caterpillars. These individuals, however, got into the habit of using those in front and those behind and walking by looping their bodies. The legs between, not being used, disappeared. The measuring worm now has no legs along its middle joints and must of necessity travel by looping.

It is incorrect to refer to any of these caterpillars as worms, but such names, having attached themselves to them and come into general usage, are hard to change.

This moth, mother of the silk caterpillar, is a domesticated creature just as is the bee or the hen. She has lived so long with man that she has adjusted herself to his requirements. Man has not, for instance, allowed her to fly. So many generations have passed without using her wings that she is now entirely unable to fly. Let the story begin with her.

The silk worm farmer selected his best cocoons as seed for next year's crop, put them by, and soon these moths forced their way out at one end. They were helpless little moths that could not fly and that seldom moved more than six inches from the spot where they were born. Through the centuries the silk farmers have been selecting the inactive breeds to reproduce themselves because they could be handled more easily, and so the silk worm

moth has become the most helpless of her race. These moths lived a few days, laid their eggs, and died. This was their province in life. It is the province of most insects during the adult stage, but most of them make a frolic of the time during which they have wings. These silk worm moths, in their helplessness, remind one of Chinese women with their feet deformed and almost useless

THE SILK WORM.

from the practice of encasing them from childhood that they may not grow.

When the mother moth is ready to lay her eggs she is placed on a sheet of muslin cloth. She lays some five hundred of them, tiny things not as big as a pin head. Then, her purpose being served, she dies.

The silk worm farmer collects the eggs. He may keep them for his own use the next spring or he may sell them to somebody else who wants to raise silk worms. You or

I, having mulberry trees in our gardens, might buy an ounce of these eggs. If allowed to have their own way they would hatch out when the warm weather of the following spring came around, but they can be held back by being kept in the refrigerator. One can keep them cool and hatch them at will by putting them in an incubator and keeping them at a given temperature just as is done with the eggs of chickens.

Naturally one would hatch his eggs at the time the mulberry leaves are plentiful. From this ounce of eggs would come no less than forty thousand tiny, black caterpillars.

These caterpillars are born hungry. Their one desire in life is to get at some mulberry leaves. There are many methods followed in feeding them, but one of the cleverest is this:

Place over the little caterpillars a sheet of paper with perforations in it through which they are barely able to crawl and with finely chopped mulberry leaves on top of it. The caterpillars immediately begin scrambling through these holes. In doing so they scrape off certain particles of egg that are likely to cling to them and would cause them to die.

They get to the chopped mulberry leaves and eat their fill. They are then allowed to get hungry again and a second sheet of paper with mulberry leaves is put over them. They scramble on to this, which is new and clean, leaving their first home, which has become quite soiled, behind them. It is burned. Thus has an ideal method of cleaning the stable for these tiny livestock been devised.

This progression upward to new food and cleanliness goes on for about a month. Each time the feeding trays have been made bigger to accommodate this growing family of forty thousand. In the end sixty times as much space is required as in the beginning. By the end of the month the caterpillar has paused four times to burst its jacket and crawl out of it, each time putting on a bigger one, as is the way of insects as they grow. Now these caterpillars are slim, hairless, pinkish-white fellows three inches long.

As they reach their full growth there is a peculiar development in their make-up. Within each side of their bodies there have been growing quite noticeable sacs or glands. Their appetites are enormous this last five days of their eating life, and it is probable that the nourishment they extract from this food goes to filling these sacs. If you should dissect the caterpillar, you would find that these sacs are filled with a glue-like liquid. It is, in fact, liquid silk. It has been extracted by one of those miracles from these mulberry leaves. It is from these tanks that the cocoon is to be spun.

Finally, its eating done, the caterpillar begins to rear its front part and reach about into space. It cannot see very well and is, in fact, feeling about for something up which it may climb in search of a suitable place to spin its cocoon. When the caterpillars begin thus to reach out, the silk farmers provide them with collections of coarse straw, twigs or lattices up which they may climb. There they find places to which they may tie threads and make a framework into which they may weave their cocoons. Their heads move about constantly and rapidly.

As they move they are pressing upon their sac, a thin column of this thick liquid is streaming from each and uniting as they issue from the creature's under jaw. When these columns strike the air, it has the effect of hardening them just as it hardens glue or chewing gum. These hardened columns of liquid become the threads of silk that the caterpillar spins.

The worm creates a cloud of silk about itself. Inside this it keeps on weaving. It does not stop for three or four days and nights. By that time the cocoon is complete and the liquid sacs are empty. The caterpillar, the larva, becomes a chrysalis or pupa. It goes soundly to sleep after all that work, confident that it will some day wake up to emerge a moth.

It does emerge a moth if the farmer happens to pick it out as the favored one that is to lay the eggs for next year's crop. The chances are, however, a thousand to one that it will meet a quite different fate, for

A SILK WORM COCOON.

the cocoons that are to furnish the silk of commerce must grip the sleeping chrysalis in the embrace of death before they can serve their purpose.

If the moth comes out, she breaks the threads and the silk will no more unwind than would a ball of twine with a hole pierced in it cutting the strands. To be unwound the cocoon must remain intact. Therefore, the creature inside must not emerge.

So these cocoons are put into trays and are run into

ovens. There the life is baked out of the chrysalis and it is thoroughly dried. All the little weavers give up their lives that the silk farmer may get his crop. Hundreds of them have died for the making of a single handkerchief; thousands for a single silken garment. To be sure, they have been pampered children, have traveled from the egg to the cocoon in the lap of luxury. To be sure, also, the journey beyond the cocoon is brief. The privilege of lopping it off is the price the silk farmer charges for making the first nine-tenths of the journey pleasant. The result of his lopping it off is that people the world around may have this delicate material as the finest of fabric from which apparel is made, as the most responsive medium for the expression of the art of the fabric weaver and printer.

CHRYSALIS SKIN INSIDE PIERCED COCOON.

These cocoons, when the time for unwinding comes, are soaked in warm water. The purpose of this is to soften the sticky substance that the caterpillar has used in binding his threads together in the walls of the cocoon. Thus softened and stirred about, the loose ends begin to appear. These must be gathered up by hand and attached to the reel which begins to wind them up. Five or six are joined to make a strand of this delicate floss. Ninety strands from ninety cocoons would need to be twisted together to make an ordinary strand of sewing silk. It takes about three thousand cocoons to produce a pound of silk, and the thread on each of these is half a mile long, and all combined would reach from Chicago to New Orleans. That ounce of eggs with which the grower started, hatching into forty thousand caterpillars, ripened, probably, thirty thousand cocoons. These cocoons would weigh about a hundred pounds. When the raw silk was finally wound on skeins it would weigh not more than ten or twelve pounds—not a weighty harvest at the end of the year. Yet there are in the world enough people caring for silk worms to produce each year nearly one hundred million pounds of raw silk.

Silk culture is carried on mostly in the homes of small farmers. They do this in addition to their other activities. It does not lend itself to large scale operation. Silk worms and white mulberry trees grow well in most parts of the United States. The industry might thrive here if great numbers of people cared to engage in it. The profits, however, are not sufficient to appeal to the American. The Oriental and the South European may attend to the infinite detail of silk culture for this small return, but not he. America's rôle in support of the industry founded by an insect is in the purchase of raw and manufactured silk, in the use of it by her people. In this she does her full share, for more than half the silk grown in all the world eventually finds its way to the United States.

QUESTIONS

1. When our ancestors were ignorant savages living in rude huts the Chinese were already highly civilized. There, in the oldest of nations, the silk industry was born. Report to your class on the following questions:

   (a) What do we owe to the keen and observing eyes of a youthful empress who lived twenty-six hundred years before Christ?

(b) What invention is she believed to have made?

(c) How long did China keep her secret? In what way did Japan get hold of the secret? What do we owe to the Persian monks sent to China by Emperor Justinian?

2. Consider the United States as a silk-producing region. Why do the Orient and Southern Europe still remain the most important sources of supply?

3. (a) You remember how the gipsy moth was introduced into the United States. With what characteristics of the moths are you familiar?

(b) Which do you consider more beauttiful—the luna or the Cecropia moth? Do you think Mother Nature's trademark adds to the beauty of the Cecropia?

(c) Is the silk worm, the army worm, and the measuring worm correctly named? Give reasons for your answer.

4. (a) What do you consider most interesting in the life history of the silk worm?

(b) What changes have been brought about in the silk worm as a result of being domesticated?

(c) How does the silk farmer feed the hungry caterpillars?

(d) Describe the larva and pupa stages.

(e) How is the cocoon prepared for commerce?

5. What is the future of the silk worm industry in China? In what way is the success of this industry dependent upon the United States?

Chapter XV

THE FLY

T HE housefly, sitting pensively there on your window, with the assistance of the members of his family who have gone before, has killed more men since time began than has any other creature. The human beings that have been slain by lions and tigers, killed by wild bulls, bitten by rattlesnakes are but a handful as compared with the multitudes that have died because of the housefly.

Man himself has been an outstanding enemy of man since the human race began. Men creatures have fought each other in private combat, in feudal conflicts, and in stupendous wars, where millions have been arrayed against one another to kill. Yet the men that have been slain by other men in all these conflicts are small in number when compared to the men that have been killed by flies.

The chief method of murder employed by this familiar creature of the household is through carrying disease germs from place to place and fastening them on people who otherwise would remain well. The existence of these diseases is due to the presence of bacteria, tiny microscopic germs of which the world was hardly aware until the present generation. Outstanding among the germ-borne diseases is typhoid fever which, down through the ages, has taken a greater toll of life than almost any other. This housefly carries typhoid fever. Possibly most of the people who have died of it since Adam have done so because the fly carried the germ from place to place.

Bacteria are so tiny that it requires a vivid imagination to conceive of their existence. By actual examination it has been shown that six million of them may attach themselves to the body of a single fly. It is no uncommon thing for the fly that comes in at your kitchen window to bring with it one million bacteria.

How, you may well ask, can anyone know that there are six million bacteria on a given housefly?

The answer is that there is a method of counting them. A scientist may capture such a fly. He may put it into a quart milk bottle that is half full of water and shake thoroughly. Thus are the germs washed off the fly and distributed through the water. Then he may take a tiny particle of this water, maybe a millionth part of it, and place it on a slide under a powerful microscope. He counts the bacteria in that particle. There are six of them. He multiplies that six by a million.

A fly carrying all these bacteria might light upon a wedge of pie which is to be served at mealtime, might wipe its feet, and in each track leave a thousand germs. These go into the system of the person who eats the pie and may there multiply to his destruction.

Conceive then the possibilities that lie in washing off bacteria from the body of the fly that may happen to alight in a pan of milk. Since the milk furnishes an excellent place for these tiny creatures to breed, it becomes an ever increasingly deadly thing for the members of the household to put into their mouths.

It is the habit of the fly to visit all about the neighborhood. Wherever unclean matter is left exposed, there flies congregate and feed. Wherever windows of kitchens are open, they enter and bring the contamination of their former journeys with them. It is thus through the ages that they have carried their messages of death.

The sanitary officers of the United States Army discovered, during the Spanish-American War, that it was the housefly that broadcast typhoid among the troops and caused the death of so many of them. With that discovery began a campaign against this little creature which, during the first two decades of the present century, carried an understanding of its frightfulness to most people and led to the world-wide campaign to reduce the numbers of the pest. The comparative effectiveness of that campaign is a tribute to the enlightenment of the people of this modern world. By the time the World War came, so thorough had the understanding of the methods of combatting the fly become that that great conflict was fought almost without the appearance of typhoid. It was the first war ever so fought. There were other elements in the situation, as, for instance, typhoid vaccination which is the demonstrated master of the disease, but the vanquishing of the fly played no mean rôle in life saving.

With this background revealing the fundamental villainy of this familiar and apparently innocent insect, we may proceed to an examination of some of its interesting traits. The fly, to begin with, is different from all other insects except those of its own order, in that it has but two wings where the others have four. This order the scientists call *Diptera*, which in Greek means "two wings." There are many sorts of flies, all of which are distinguished by this peculiarity. The gnats which fill the air in the summer-time, for instance, are but tiny flies. The blue-bottle fly, which bumps its head so vigorously against your window pane, is but another variety. The horsefly, that finds itself a place upon the back of Dobbin out of reach of the switch of his tail, is another sort of fly. The mosquito, which buzzes so disquietingly of a summer evening, belongs to the same order of two wings, but its relationship is a bit more distant.

A generation ago a popular query often propounded in theatres and humorous journals was this: "Where do the flies go in winter?" A serious answer to it later became a matter of great importance in the campaign for the suppression of the fly. It became known, as a matter of fact, that most of the flies die in winter, but it is likewise certain that some few of them find places of warmth and protection, live through the cold season, and start new generations in the spring. If these few could be killed, the millions that follow them would not be born. This idea is the basis for the campaign for killing flies in the spring. One fly killed in the spring may mean millions fewer in August.

The rapidity with which flies multiply is one of the marvels of insect life. It has been shown, for instance, that in the summer-time when the generations follow each other most rapidly, one fly may become millions in a few weeks. A female fly on a summer day may lay 120 eggs. She may lay four such hatches, so figuring on a basis of 120 is conservative. These eggs hatch into larvæ in eight hours. These larvæ grow and eat and thrive and in five days harden into chrysalis state, there to change themselves from worms to flies. They remain in this state for another five days and come out as adults. Thus, in ten days has one fly become 120, of which 60 are females. These females are mature for ten days before they in turn lay 120 eggs each. So from egg to egg twenty days have elapsed. In the third generation there are three thousand six hundred females to lay 120 eggs each, resulting twenty days later in 216,000 to do the same, and in another twenty days in 12,960,000, and so on. This schedule, begun in April and continued to September, might result, theoretically, in the presence in the world of some six trillion flies because of the one individual that lived through the winter. Thus, if it were not for accident in birth and development, might a city like Washington find itself

EGGS OF THE HOUSEFLY.

in September completely buried beneath the mass of flies that came from a single mother in April.

One of the discoveries of the scientists, which came to pass when a serious study of the housefly was undertaken, was the fact that ninety per cent. of all of them breed in manure piles adjacent to stables. The flies lay their eggs in these manure piles, which furnish abundant food for the larvæ and final hatching places for the adult fly. The fly population of the United States has been very greatly reduced since the beginning of the present century, and its reduction has been largely due to the elimination of the manure pile. The present generation is hardly aware of the fact that in villages and farmhouses no longer ago than 1900 it was the familiar thing to find screen doors entirely covered with houseflies, and the battle against these pests one of the chief summer duties of the housewife. The elimination of or treatment of the manure pile has almost removed this erstwhile conflict.

Yet this fly, which lives wherever men do all around the world, and is so familiar that nobody pays any attention to it, is not an uninteresting sort of creature and has some accomplishments of which even man might well envy him. It is a rather remarkable thing, for instance, that it should be able to take its promenade up the slippery side of the most highly polished pane of glass or stroll with unconcern head down on the ceiling.

The fly, that it might accomplish these feats, has spent many thousands of years in developing for itself a very specialized sort of foot. On each of these feet it has two pads like the fleshy part of your hand. On each of these pads there are hundreds of tiny, hollow hairs with openings in them much like those of the rubber spray brushes for the bath which you see in the drugstore window. Each foot of the fly is connected with a pot of glue, and, when it feels itself in danger of slipping, it squeezes a bulb and the glue comes just to the surface in these openings in its foot pads. It sticks them to the window pane. The walking is a bit heavy, something like that of plodding through thick mud, but is better than slipping to a fall.

It has long been said that houseflies bite when it is

about to rain, that one could sit inside his own home and they would warn him of the coming of a storm. This observation has been repeated for centuries. Then along came the modern scientist and established the fact that houseflies do not bite at all. They cannot with the sort

WHERE THE HOUSEFLY BREEDS.

of mouth they have. They are suckers and not biters. Their tongue is a sort of sponge, abundantly supplied with saliva when they want to dissolve sugar, and back of it a suction pump. It is thus that they eat, dissolving their food and sucking it into their stomachs. Yet, despite this, they bite ahead of a storm.

A solution of the riddle of how this could be possible was quite difficult. Finally it was shown that the stable fly, very like the housefly in appearance, did bite. The stable fly, it was also shown, went inside when it looked like rain. Once inside it set about biting human beings just as it would bite the horses in the barn.

This stable fly is also a creature of frightfulness. It is suspected that it is the visit of the stable fly and its biting that carries infantile paralysis to many children. Hundreds of thousands of human beings may have gone through life as cripples because of this stable fly.

Yet this fly has helped make American history. There is no date more outstanding in the calendar of the United States than July 4, 1776. It was on that day that the group of men which framed the Declaration of Independence, sitting there in Independence Hall, in Philadelphia, were wrangling without end, as is the way of such groups, over some detail of the document. Though history does not say so, it must have been a sultry day with a threat of approaching rain. It is set down that there were barns near by and that innumerable flies came in by the open windows. These were unquestionably stable flies, which would not have been inside if a storm had not been in the offing, and not houseflies, for they pierced the silk stockings on the calves of the fathers, and bit them until they were exasperated beyond measure. Finally one of them, in desperation, proposed that they sign the Declaration forthwith and escape these abominable flies. All agreed. So was the act of signing actually brought about.

The fly is remarkable as an insect for the perfection of

its arrangement for seeing. Its eyes are among the most highly developed of those of any of the insects. It has more of them than most. It can actually see as much as two and a half feet, which is better than the majority of insects can do.

The fly has five eyes, two of which, known as compound eyes, are organs incomparably more complicated than are those of man. These two amber, major eyes are very conspicuous in the fly. They occupy most of the front part of the head. If our eyes were as big in proportion to our size as these major eyes of the fly, they would stand out as great projections much larger than our fists.

Then there are the three shining, simple, skylight eyes of the fly, set in a neat triangle on the top of its head. These simple eyes are little more than windows to let in the light. They are composed of a single lens which is made up of chitin, the material of which the skin, or outside skeleton of the insect, is composed. This eye is a convex lens. The surfaces on each side curve outward. Below it there is a retina which receives the sight impression and an optic nerve that runs to the brain and carries the message of what is seen.

Scientists are able to measure the focus of these simple eyes of the insect just as they are able to measure the focus of the lens of a microscope. They find that the focus is so short that an insect could not see with them for a greater distance than one inch. Such insects as bees and wasps that live in narrow and dark quarters are likely to have these simple eyes more highly developed than have others. Some night-flying insects have only these simple eyes and it is therefore assured that they see but little.   Butterflies, which go out by day, have no simple eyes, but highly developed compound eyes.

The excellence of the vision of the fly, which is shown by its quickness in escaping, if attempts are made to catch it, is due to the presence of its big compound eyes.   These are also present and highly developed in such other creatures as butterflies and dragon-flies.   The compound eye is built on the same principle as the simple eye.   It is composed of a modified skin, of chitin, which hardens into

THE STABLE FLY.

a lens and back of which there is a retina and an optic nerve.   It is, in fact, a group of simple eyes packed close together.

The outer convex lens appears on the surface.   There is no lid to the insect eye and its surface requires no moistening as does that of the eyes of more highly developed animals. The fly cannot close its eyes and, further, they are stationary, and it cannot turn them in one direction or another.   It is because of this fact that many lenses are needed, each set at a different angle so that the fly may see all around without the necessity of turning its head.

It is surprising to find the great number of lenses in these compound eyes.   The simplest of them are to be found in a Brazilian beetle, the eyes of which have but seven facets.   The eye of the ant, which sees very little, has fifty facets.   Much more complicated is the eye of the housefly, which has 4,000 facets.   This is far from being a record of separate lenses in a given eye, however, as those of the swallow-tailed butterfly have 17,000, while those of the dragon-fly have 20,000. The record is held by the eye of a certain hawk moth, which runs as high as 27,000 lenses.

These lenses are six-sided and, when viewed from the front with a powerful microscope, look not unlike the surface of a piece of honey comb. They are packed together in the same way as the cells of a honey comb and tie back, of course, to a common retina and optic nerve.

Despite this very complicated arrangement of the insect's eye, it is, after all, a quite ineffective organ of sight when compared to the eye of man. Certain butterflies are thought to have the best eyes in the insect world, and yet they are barely capable of seeing the movements of large objects at a distance of five feet. The range of visions of bees and wasps is generally not more than two feet. The housefly cannot discover the fact that attempts are being made to "swat" it until its pursuer is within two and one half feet of it. Thus does it appear that, while the sense of smell is much more highly developed in insects than in the larger animals, the sense of sight is quite rudimentary.

The blue-bottle fly makes quite a racket with its wings because they move so rapidly. Scientists, always hungering to isolate new facts against the possibility of some use being found for them, have tried to discover the number of the vibrations per second of the wings of the blue-bottle fly. These vibrations come too fast to count, but they create a musical tone as does the tuning-fork, the vibrations of which are known. The pitch of the tuning-fork depends on the rapidity of its vibrations. The pitch of the tone of the wings of the blue-bottle fly has been compared with it, and it has been shown that they vibrate 300 times a second, which is going with considerable speed.

The blue-bottle fly, however, is one of those few creatures in the insect world that can make a noise other than with its wings. In forcing the air in and out of the slits in its sides through which it breathes, it can produce that well-known drone which everybody has heard.

This blue-bottle fly is the scavenger, the buzzard, of the insect world. Its larvæ live in meat, devour it, cause it to decay, and get out of the way. Most flies are, in fact, scavengers. They and their young eat up unclean objects. In this way they do a certain amount of good in the world to pay in part for the havoc they work in carrying disease.

They are, however, nearly always deadly. There is the buffalo gnat, for instance, that tiny hump-backed fly which tortures cattle and broadcasts disease among them. In Africa it is the tsetse fly that carries the dread sleeping sickness to its victims. Flies are responsible for the summer complaint of babies.

This two-winged breed is, in its relations to human beings, the outstanding exponent of frightfulness in the insect world. Through the centuries it has slain man and beast in countless numbers. Yet man himself has not known until the present generation the source of the slow death that was being dealt out to him. The murderer fly killed him as might a villainous guest in his house who secretly put poison in his food. By the beginning of the present century he had discovered the treachery of this guest and he then took up the age-long challenge of Diptera, called the members of his own tribe everywhere to recognize this, their ancestral foe, as the murderer he was, to exercise eternal vigilance and to give no quarter in the war of its extermination.

QUESTIONS

1. (a) We often think of wars causing much needless loss of life.
   (b) What enemy, more deadly to man than war, is still in our midst?
   (c) Now that you have read the story will you join the "Swat-the-fly" movement? Why?
   (d) How can little boys and girls help?
   (e) What can housekeepers do toward exterminating the common housefly?
   (f) Compare number of flies found in the farmyard to the number two decades ago.
2. (a) Science has done much to make this world a safer place in which man might live. What has it discovered about typhoid fever?
   (b) What precautions were taken to eliminate typhoid fever in the World War by our government? Compare the sanitary conditions of the World War with those of the Spanish American War.
   (c) What havoc is brought about by the tsetse fly in Africa?
3. Suppose you had as a neighbor an ignorant person who left his garbage can uncovered, who did not believe in screening his house, who allowed flies to settle on food prepared for meals, how would you proceed to make clear to him the dangers flies may cause?
4. (a) Have you ever observed a fly walking on the window pane or on a ceiling? How has nature equipped the fly to perform this feat?
   (b) What part of the fly's head helps him in making his escape?
   (c) Compare the vision of the housefly with that of the butterfly and other insects.
   (d) Illustrate the part played by the horse fly in our American history, July 4, 1776.

(e) Illustrate the part played by the blue-bottle fly, the buzzard of the insect world.

5. How many species are there in Diptera, the order of flies? Resolved —that the housefly, sometimes called the typhoid fly, is not only an annoyance but a positive danger.

Chapter XVI

THE MOSQUITO

THERE is no natural reason why a mosquito should bite a human being. Not one in a hundred million that come into the world ever does so— ever has an opportunity to do so. The natural home of the mosquito is the swamp, the wastes of the tropics, the great stretches of such regions of the north as interior Alaska. There are few men in these places. Not one mosquito in a million ever sees a man. They run the courses of their lives in a quite natural way without doing so.

Then it happens that an isolated mosquito comes within range of a human being. That mosquito and all its ancestors back to the Adam of its family have gone along peacefully feeding on plants with but an occasional nip at some animal of the woods. There is some instinct buried within it, however, that causes its nature to change when it smells a man. Its instincts change from those of a plant feeder to those of a blood drinker. It becomes consumed by a single purpose, that of driving its drill into this human's skin, setting its pumps to work and filling itself to bursting with his blood.

It is only the female mosquitoes that are thus blood-thirsty. The males are harmless because they have no effective drills. They may fly around with their plumed antennæ giving all the appearance of full beards, thus distinguishing them from the unwhiskered women, but they cannot bite. Give the female a chance, however, and she will carefully select a spot for drilling, will quickly penetrate to the red, and will begin to gorge herself. They have been seen to drink blood until they actually burst. Even though they do not go that far they take aboard such a cargo that their wings can scarcely carry it. Watch them making their retreats and you will observe their labored flight and the fact that they must stop for a rest every few feet. If one resents the attack of the mosquito and crushes it while it is in the course of getting its meal, the result is likely to be a splotch of blood as the chief evidence of the wreck of its frail form. It should be borne in mind that this blood is not that which flowed in the veins of the mosquito, but that which it had drunk from its victim. The blood of this mosquito, like the blood of all other members of the insect world, it should be borne in mind, is not red, but pale and colorless.

This blood-thirsty female, when she drilled her well into the human blood spring, used a quite complicated instrument. It was composed, in fact, of four drills within a sheath. When she had reached the blood supply, she found it advisable, that she might drink freely, to thin it a bit. She therefore squeezed a bulb which contained her reserve supplies of saliva and thereby injected this liquid into the wound and it spread through the blood near it. Her drink being thus thinned in accordance with her taste, she reversed her pumps and the blood began to flow into her own body.

When the wound received from the bite of the mosquito tends to smart and swell, it is because of the presence in it of this liquid that the visitor has injected. If the mosquito is allowed to sit quietly and drink its fill, it will withdraw with its meal the greater part of this liquid, and the result will be that there will be little annoyance from, or swelling of, the wound. If, on the contrary, the mosquito is disturbed early in its meal, is killed or driven away, the somewhat poisonous liquid remains and inflammation follows. It would seem therefore that the logical advice to one who is being bitten by a mosquito is to exercise the instincts of hospitality and patience, to allow the visitor to complete its repast.

As the possession of these drills with which she may penetrate the skin is an outstanding peculiarity of the female, so is the possession of fluffy, beard-like antennæ, constructed very much on the plan of the ostrich plume, a distinguishing feature of the males.

The plumes of the male mosquito, however, serve a very definite and highly scientific purpose. Of late years it has come to pass that, when a steamship enters the port of New York, it begins to listen for a signal beneath the surface of the water. It has very delicate instruments on each side of its hull which it uses for picking up this signal. When it hears the signal on the port side, it realizes that the sending station is on that side. The nose of the ship is turned in that direction until, finally, the signal is picked up by the instrument on the starboard side. When the instruments on both sides of the ship are getting the signal in the same degree, it is known that the nose of the ship is turned directly toward the sending station. That sending station is at the New York end of the ship channel and, by going directly toward it, the vessel is certain to avoid shallow water.

So, in the case of the male mosquito, experimenters have found that, when a certain note is sounded on a tuning-fork, the plumes on one side of its antennæ may begin to vibrate. The mosquito receiving the message turns about, facing the sending station until the plumes on its antennæ on the other side vibrate in the same way.

A MOSQUITO EGG.

At this point it knows that it is head on toward the signal it is receiving. This note to which the male mosquito seeks to respond is the voice of the female mosquito, that song which harasses millions of human beings all around the world every summer evening. The insect makes it, as the blue-bottle fly does its drone, by blowing through the air holes in its sides. When the male mosquito goes to meet this sound he is, in reality, seeking a mate.

One observer of mosquitoes found great numbers of them lying dead beneath a certain street light each morning. Upon investigation he discovered, oddly, that all of these mosquitoes were males. Further investigation revealed the fact that this light in its burning made a noise, struck a note which corresponds to this signal of the mosquito world. It vibrated a call to all the male mosquitoes in the vicinity and, determining its direction by their antennæ, they flew to it, while the female mosquitoes, without receiving stations keyed to this note, failed to respond.

The mosquito, it should be borne in mind, is a cousin to the fly. It is a member of the "two-winged" order, *Diptera*. In Spanish, "mosco" means "fly" and the mosquito is therefore "the little fly." The old English word for the little fly is "gnat." In England the terms "gnat" and "mosquito" are interchangeable, but in America they are applied to different but related families of the order Diptera.

The mosquito offers itself as one of the most convenient families of insects to be studied. Unlike the other flies, unlike even the gnat, it is born in water. It breeds almost anywhere that still water is to be found. Swamps are its natural habitat. Standing pools of stagnant water anywhere serve its purpose. There is no better breeding place for it than the old-fashioned water barrel which sits under the eaves to catch soft water when it rains hard. A tin can in a rubbish heap which catches a bit of water becomes an incubator for baby mosquitoes. A discarded bottle with water in it may breed thousands of them.

Whoever wishes to observe the life cycle of the mosquito may do so by dipping up in the summer-time a milk bottle half full of water from any stagnant pool. If the mosquito has not already laid eggs in this water, she will probably appear soon to do so. The tiny eggs of some mosquitoes sink to the bottom and there hatch. Most of them, however, bind together their clusters of eggs, several hundred in number, in a tiny raft, which they fit with air bubbles to keep it afloat. Presently the eggs in this raft hatch out, and the youngsters plunge headlong into the water. It is a joyous plunge, however, because the mosquito, unlike its fly cousins, starts out by being a water animal.

These mosquitoes in the larval stage are the familiar "wiggle tails," the actions of which are known to every child. They are odd and active little fellows trimmed in bristles, and with the heads much bigger than their bodies. One of their most interesting feats is the manner in which they get air to breathe, for they are not wholly water creatures and would drown if kept too long beneath the surface. On the second joint from their tails these wigglers

MOSQUITO LARVA.

have what would appear to be a periscope, sometimes half as long as their bodies, just like those the submarines ran to the surface and peeped through during the war. It is not a periscope, however, because it is not made for purposes of observation. It is a ventilation pipe. The wigglers approach the surface of the water and send up this pipe. There is a float on the end of it which brings it just to the surface. The wigglers breathe through it and remain there at anchor, as it were, head down, taking their rest.

An observer of these wigglers will see that, in the course of their growth, they shed their skins three or four times, on each occasion getting a bigger jacket, as is the way of the insect world. Finally they have reached the period of development where they should become pupæ, enter that stage in which most insects hibernate, get a long sleep while changing from the grub form to that of a flying creature of the air. The mosquito pupa, however, refuses to go to sleep, although it does stop eating. If you watch it in your milk bottle, you will find that, instead of becoming a motionless pupa, it develops into an odd form, the like of which is not to be found anywhere else in nature. This mosquito pupa becomes an awkward, roundish creature, appearing to be mostly head, but with a tail-like abdominal section. It is shaped for all the world like a comma. It now has two ventilator tubes sticking up like donkey ears. At the end of its abdominal section there are two fluttering, leaf-like appendages which it uses for purposes of navigation. When it flutters them, this tail of the comma naturally pushes on the side of the bigger roundish body of the creature to which it is attached. The result is that the whole pupa turns over and over in the water and its method of travel is by means of this somewhat comic system of rolling.

Inside this awkward pupa the delicate form of the mosquito is taking shape and can be observed to some extent through its shell.

Then comes the time when the mature mosquito is to be born. The pupa lies at the surface of the water. Its skin cracks open down the back and the dainty mosquito emerges. It perches upon the habitation it has just left, using it as a raft until the chitin of its outside skeleton may harden and stiffen and its wings may dry that it can take flight. This moment upon the raft, while it is yet helpless as a creature of the air, is a precarious one for the mosquito. It often happens that it loses its footing, is blown over by a rough wind or otherwise thrust into the water from which it has so recently emerged and, now being a creature of the air, there drowns.

This mosquito, down through the ages, has been an unceasing irritant to man, has broken his rest, has disturbed his sleep, has caused him to expend much energy in fanning at hands and face and ankles in dislodging an uninvited guest. Man regarded the mosquito merely as an annoyance until, during the present generation, scientists went into the realm of infinitely small creatures, so small that they consist of but a single cell, and there traced a life cycle which revealed the manner in which certain death-dealing diseases which have killed millions were developed, later to be broadcasted by the mosquito.

These scientists found that the mosquito, belonging as it does to the order of flies which is so malignant, so deadly, as a carrier of typhoid and other diseases, serves a no less tragic purpose. The mosquito has been shown to be a bearer of two particularly deadly diseases, malaria and yellow fever.

A one-celled animal which the mosquito carries about is the cause of malaria. This infinitesimally small creature, known as a protozoan, is first discovered as a mere speck in a single red corpuscle of the blood. Only very powerful microscopes can see it. It is, however, a living animal just as is an elephant. It demonstrates this fact by setting to work to eat that corpuscle, just as the grub of the parasitic wasp eats the moth egg in which it is deposited, or just as the donkey eats the hay in its manger. When it has devoured the contents of this corpuscle, it does a thing not encountered among larger animals.

It splits up, and each fragment of it sets out to find a new corpuscle to devour. In the course of this action these tiny animals release toxins or poisons. These probably come from a lack of proper arrangement for garbage disposal in their households. It is these toxins that cause people afflicted with malaria to have chills.

Later in the development of the protozoa in the human blood, they take the form of wigglers, and in that stage find their mates. This, however, is the end for this tiny creature unless it can find the peculiar incubator which it requires for developing others of its kind. That incubator does not exist in the human body. It is found only in the stomach of the malarial mosquito. Unless this protozoan can escape this human world in which it has grown up, it can never hope to breed other generations of its kind. Its race, unless it does escape, must die. There is but one means of escape and that is through the bill of a malarial mosquito. If this peculiar sort of mosquito visits the human being who is suffering from malaria and sucks his blood, it will, in that blood, acquire some of these protozoa. It is the only chance of life for them.

The protozoan develops within the mosquito and takes on a different form. One of them becomes 10,000 within the body of their host.

This mosquito, thus ateem with this tiny creature which breeds malarial ague in human beings, then fares forth in search of a victim. The protozoa in her system were received from one human being, and she will pass them back to another in improved form and in good measure. In the night she prowls about, singing her monotonous song, finds a sleeping victim, drills her hole, injects her saliva and, along with it, an enemy army which will attack the red corpuscles of his blood. So is malaria incubated and transmitted.

The understanding of the manner in which yellow fever is transmitted grew out of previous discoveries with relation to malaria. The fact that yellow fever is transmitted by a mosquito and in no other way was established during and after the Spanish-American War which carried American troops and sanitary corps into the tropics and exposed them to this dread complaint. The discovery of the manner in which yellow fever is transmitted by a mosquito is regarded as one of outstanding importance in the history of disease.

This, the deadly plague of the tropics, which used to be malignant in such cities as Havana and, upon occasion, found its way into New Orleans and even Philadelphia, has now largely ceased to exist. It formerly ran riot because of the activities of a certain little mosquito well known in the tropics, a domesticated mosquito which lived in houses and operated in broad daylight, a mosquito that knew the ways of man and how to circumvent him. The yellow fever mosquito was not so foolish as to betray its presence by its song. Being domesticated as it is, the noisy kind have been exterminated long ago. But, unlike the malaria mosquito, it must have a diet of blood to get along. And in getting that blood it carried the germ of yellow fever from one person to another.

Man, knowing that yellow fever and malaria were due to the presence of the mosquito, has now directed his energies to a control of that insect and, by making that control effective, has reduced the former to an impressive minimum, and produced a very material effect upon the prevalence of the latter.

A MOSQUITO BREEDING PLACE.

The general principles in combatting mosquitoes are very simple. The first thing to do is to see to it that there is no standing water about the place. Pools should be drained or filled in. No water should be allowed to stand in discarded or other vessels. The rain-barrel, if one is necessary, should be screened. Wherever it is not possible to drain standing water, it may be made mosquito proof by the simple expedient of pouring a small amount of kerosene upon it. This oil spreads out in a thin film upon the surface and, when the mosquito larva or pupa comes up to breathe, it gets oil instead of air and the result is fatal. A demonstration of this may be given by putting but a teaspoonful of kerosene in the milk bottle which you have been using for the observation of the life of wigglers. Very soon they will all be dead.

Man's enemies of the insect world are numerous and deadly. The conflicts between man and certain of the insects are among the most desperate fights that are going on in the world today. When a rattlesnake bites a hunter in the woods and he dies, the excitement is great. But every day more deadly attacks are being made by these members of the fly order of insects on tens of thousands of men and babies, and it is but recently that we have begun to fight back. Man, finally, has learned that the order of "two-wings" must be marked for destruction and that every intelligent individual must be depended upon to do his duty in the circumstances. It is a fight for self-preservation.

QUESTIONS

1. Of which sex of mosquito was the poet Bryant thinking when he wrote:
   "Thou'rt welcome to the town; but why come here
   To bleed a fellow poet, gaunt like thee?
   Alas, the little blood I have is dear,
   And thin will be the banquet drawn from me."

2. Topics for special reports:
    (a) The natural home of the mosquito.
    (b) The breeding place of mosquitoes.
    (c) The life history of the mosquito—larva, pupa, adult stages.
    (d) Cause of itching and swelling that follows a mosquito bite.
    (e) How the male mosquito attracts his mate.
    (f) The mosquito as a carrier of malarial and yellow fever germs.
    (g) Communities that have successfully fought mosquitoes—Havana, Panama Canal Zone, New Orleans.
3. (a) How does the malaria germ enter the body?
    (b) What causes the chill in malaria?
    (c) What effective cure have we for the malaria germ?

VOLUNTEER WORK

4. (a) Describe the remarkable and heroic experiment carried on by Dr. Walter Reed.
    (b) On the tablet marking the grave of Dr. Reed is this inscription: "He gave to man control over that dreadful scourge, yellow fever." What other stories do you know proving the heroism of the United States army surgeons and other self-sacrificing scientists? Pamphlets on this subject may be obtained from the United States government.

DISCUSS THESE TOPICS

5. (a) Prevention is easier than cure and far cheaper: therefore, it pays to make all standing water mosquito proof.
    (b) Laws against breeding mosquitoes are sadly needed in many places.

Chapter XVII

THE BORING BEETLES

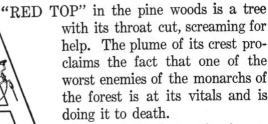 "RED TOP" in the pine woods is a tree with its throat cut, screaming for help. The plume of its crest proclaims the fact that one of the worst enemies of the monarchs of the forest is at its vitals and is doing it to death.

This enemy, but a tiny insect, in the course of getting its breakfast is wastefully killing this towering tree. It is cutting mere thimblefuls of its food from places so vital to the life of the tree that the wounds it inflicts are fatal.

The pine beetle, a blackish little creature no bigger than a grain of rice, is at work just beneath the tree's bark. Perhaps it is starting a plague of beetles that will run through the forest as disastrously as might a roaring fire.

Already it is too late to save the life of this scarlet tree. It is too late even to punish the murderer, for he has committed his crime and made his getaway. The wise policeman of the forest tries to detect its tragedies in their earlier stages when there are better opportunities for apprehending the criminal than in the case of the "red top." The pine tree first gives evidence of the attack upon it by showing a paler green than that of its healthy fellows, a green which becomes yellowish, turning brown. During this development toward the "red top" the murderer is still on the job and may be caught and punished. The interest of the community requires his punishment else his tribe may multiply and change this verdant forest into a waste of ghostlike trunks of dead trees.

This murderous pine beetle flies about the forest through the summer on its clumsy sheathed wings. The purpose of its flight is to select new tree victims. Settling on a given pine as its haven, it drills through the outer bark and into that portion of it that is next to the body of the tree. There it begins cutting its tunnels and laying its eggs. Unfortunately this is at exactly the depth at which the sap is flowing up from the roots to nourish the tree. When one of these tunnels runs across the grain of the wood it cuts all these sap arteries. When enough of these are cut the tree can no longer get its nourishment and dies.

In the meantime the eggs have hatched, have become tunneling grubs that help in the work of destruction, have developed into the transformation stage and gone to sleep in the outer bark. There they wake up as beetles and fly away to start attacks on other trees.

The modern forester fights these tiny tree killers. During the winter time, when the development of the pine beetle is very slow, he goes into the woods, locates the trees that are paling from the attacks, that are infested, and cuts them down. If there are streams available, he throws these tree trunks into them, thus drowning the beetle larvæ that are still asleep in the bark. If there are no such streams, he rips off the bark and burns it. Thus does he reduce the number of pine beetles which, in the spring, would go forth to kill other trees.

This wounded and dying tree becomes a favorite abiding place for another sort of beetle, the ambrosia beetle, wielder of a first class auger and founder of a peculiar civilization which exists deep in the wood of trees and logs that are wounded or felled and in which the sap is in a state of fermentation. Their taste runs to dying timber.

The oak shelf in your bookcase, the leaf which you

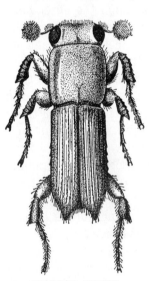

put into the extension table, the plank that you may have encountered here and there in your experience that has been full of peculiar perforations, of tiny holes bored miscellaneously through it —these are former homes of the ambrosia beetle. These holes were put there by this beetle which lived for a while within the log from which these boards were sawed, usually after the tree was cut, yet before it had become seasoned.

The ambrosia beetle lives in the sapwood between the bark and the heart of the tree. It lives there during the time that

AMBROSIA BEETLE.

the log is dying because it is then that conditions exist that are just suitable to its remarkable skill as a gardener and its scheme of growing its own food here in the body of the log. It lives by gardening, by raising ambrosia. It is from this ambrosia that it gets its name.

This food is made up of fungus which grows inside the

tunnels and excavations made by this little beetle, a quarter of an inch in length. This fungus and the manner of its growing is for all the world like asparagus. These beetles cultivate their garden under these peculiar conditions, proceeding much as might the man creatures who grow their crops in the sunlight outside. They prepare their asparagus beds, fertilize them liberally with manure, plant their seed.

This beetle asparagus, like that of man, must be cut at a certain stage of its development to furnish satisfactory food. If so cut, it will sprout again and grow crop after crop from the old stalks as does the sugar cane of Cuba.

This fungus of the tunnels, however, is a menace to their inhabitants as well as their chief support. If they plant more of it than they are able to harvest, it grows rankly, goes to seed, fills up and obstructs their chambers, smothers them. Thus it is imperative that the ambrosia beetles exercise careful judgment in the planting of their crops.

It is the fermenting sap of the recently living log that nourishes these crops. A time is sure to come when these logs will be too dry to serve their purpose. The female beetles, more alert, probably because theirs is the responsibility for the next generation, know this, and leave the drying log in advance of the day of shortage. They go forth and seek logs fresh felled, or standing trees that have been killed by the girding beetles, and there establish new homes. The male beetles remain in their comfortable quarters. With half the population gone, these males soon find themselves unable to keep the asparagus beds cut. The vegetation slowly encroaches and overwhelms these males. Often it is found that groups of them club together in certain tunnels and fight the oncoming jungle, surviving as long as they can. In the end, however, they are overcome, and may often be found huddled together in their last stand.

Another boring insect that one is likely to encounter now and again is the powder-post beetle, tiny shred of a creature which attacks the sapwood of such hard timber as hickory, ash, and oak after they are dried and seasoned, presenting stubborn surfaces to the carpenter's nail or saw. They may give evidence of their presence in the dust from borings beneath one's office desk. They may work in the handles of the tools there in the wood shed. The very timbers of the house in which one lives may become infested by them. They do great damage to stocks of seasoned wood that may be held for use by manufacturers. It was these powder-post beetles which some years ago attacked the beams of Westminster Abbey and threatened its destruction. They are hard to combat but, poison solutions put on like paint with brushes is often effective. At Westminster, however, the beams were so weakened that it was necessary to replace them.

There is probably no member of the insect world that has established itself a better known type among human beings than has the bookworm. The human bookworm, given to boring his way between musty covers, is to be found in every library, in almost every family. It is an odd fact, however, that the original insect bookworm from which this student takes his name has almost ceased to exist. It is more a historical than a literal character in the family of insects. The scientists, however, know a good deal about it. They know, for instance, that it is the young, the larva, of a certain sort of tiny beetle. This beetle used to lay its eggs among damp and musty books, where they hatched and immediately set about driving their tunnels. The possibilities that lay within the drill operated by this small creature are nothing less than marvellous. There is a record, for instance, of thirty-seven volumes standing side by side on a book shelf which were penetrated by one small worm. It drove its tunnel through them, regardless of book covers, introductions, dulness of text, details of appendices or aught else.

The ravages of the bookworm are to be found in many ancient manuscripts, particularly those which have come from warm and damp climates. In modern libraries the bookworm hardly exists at all. In the Congressional Library, at Washington, one has never been seen. The care taken of the modern library is unfavorable to their existence, and the modern publisher has a disconcerting way of putting a bit of insect poison into the glues he makes for bindings that is very discouraging to insects.

King of all the boring insects is the Capricorn beetle, or goat beetle, so called because of the length of its horns that resemble those animals. It is born to live the greater part of its life in the larval stage there in the heart of the great oak tree. Hatching from its mother's egg, it digs in deep and begins three years of precarious wanderings through the solid wood which furnishes it both shelter and food. The tunnel it makes in the beginning would fit a slender straw· As the grub goes on and on, however, its dimensions increase, and when, after three years in its wooden solitude, it prepares to emerge, it has become a stout grub as big as one's finger and very tempting to the palate of the woodpecker which goes about thumping the trunks of trees, making soundings in its attempt to locate these grubs.

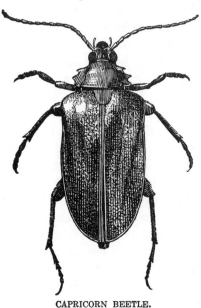

CAPRICORN BEETLE.

This grub of the capricorn beetle is little more than a scrap of intestine, a delicate bit of a grub with a white skin as soft as satin. Here in its lightless prison it has no use for eyes and does not develop them. Neither does it need ears with which to hear, nose with which to smell, nor even taste with which to discriminate its food, for there is no choice in food. To this grub, therefore, there is little left other than the single sense of feeling.

So frail a thing it is that the wonder is great that it is able to drill this tunnel through wood which would test the stoutest auger. The tools with which it does this work are spoons with a razor edge attached to its jaws, its mandibles. With these it scrapes incessantly, shaving off the surface that is before it, extracting the juices from its shavings for its own nourishment and piling its rubbish behind it. Each day it has gone a little way on its three-year journey.

To operate an auger like this there must surely be something against which the insect braces itself to exert its pressure. Yet it has not even a foot to put to the ground for a push. It has no legs upon which it can walk. Lay it there on your desk and it cannot crawl an inch. It can more easily bore through a board than crawl across it.

This grub of the Capricorn beetle in fact is constructed to walk in a tunnel. In doing this it uses one of the oddest devices in the world as a substitute for feet. These are rings about its body that can be brought to bear as feet above, below, or on the sides. These rings can be swelled to fill the tunnel or slackened to pass through it at its will. Wanting to move in this tunnel, this grub may extend the rings on the back part of its body until they fill it and take hold of its walls. Then the grub can stretch out its body to its full length ahead. Then it can expand the front rings, swell up its muscles until they take hold. The back rings are then slackened and drawn up. They get another hold and press the body forward again. Thus does it make its own slow progress in its peculiar habitat. Thus does it grip the walls when it would press upon the auger with which it bores.

The great day for the Capricorn larva comes after three years of boring when an instinct tells it that soon it is to enter the long sleep from which it will emerge a strange and compelling creature, an armored and long-horned beetle. From its place there in the depth of the tree trunk, it turns its drill toward the light of day and bores steadily, possibly through half a foot of hard wood. Finally a point is reached where the feel of its drill tells it that it has almost reached the outside. Here it stops with a layer of wood as thin as a piece of paper before it and sets about making elaborate preparations for the birth of the beetle that is to be.

Going a little back into its tunnel, the grub excavates the transformation chamber. This is two or three inches long and wider than it is high—such a chamber as this larva has never dug before. From its walls it scrapes infinitely fine shavings and lines this chamber with them until it is padded as soft as velvet. Then across the mouth of the chamber it builds a barricade which it seals on the inside with a lid made of a chalk-like substance which it secretes from within itself. With everything thus in readiness for the unhardened beetle that will come from the cocoon after the sleep, it lies down to wait the transformation.

One other precaution it always takes. In this transformation chamber it, as a grub, has had no difficulty in turning about, in going which ever way it willed. Instinctively it knows, however, that when this beetle is born it will be a creature of stiff armor plate as incapable of turning about in these narrow quarters as is an automobile in an alley. There is consequently one vital thing to be borne in mind when the grub lies down to sleep. Its head must be toward the door. In all the history of Capricorn beetle investigations no case has ever been discovered in which this humble grub with but a single sense neglected to take this precaution.

QUESTIONS

1. When we look at the magnificent pine trees stretching their giant branches to the sky, we cannot help noting their strength. It is hard to conceive how a tiny insect can cause the death of such as these. And yet this is so! Why are the wounds made by the pine beetle fatal to this monarch of the forest?

2. (a) What tells the forester of the presence of the enemy at work?
   (b) How does he fight these tiny tree killers?
   (c) Note the precautions taken by the careful forester in winter to reduce the number of pine beetles in the spring.

3. (a) We generally think of live, growing things as being objects of attack from enemies. The ambrosia beetle is an exception to this general rule. Note its home, method of getting a livelihood, and its gardening abilities.
   (b) In what way may the "beetle asparagus" prove a menace to the inhabitants?
   (c) What precaution does the wise mother beetle make in selecting her home? Note the fate of the males.

4. (a) The boring beetles are very destructive. How did the "powder-post" beetle threaten one of the world's historic places of interest?
   (b) Examine the hardwood furniture in your home. Do you find any evidence of this beetle's former attacks?
   (c) How did the human bookworm get its name? Why are real insect bookworms less common today?

5. The capricorn or goat beetle is an interesting example of adaptation to environment. In illustrating this law, consider the following factors:
   (a) Lack of eyes, ears, and sense of taste.
   (b) How fitted for tunnel life.
   (c) Means of drilling its tunnel.

6. We have read of wonderful and magic-like happenings in the insect world. What interested you most in this story?

Chapter XVIII

THE BOLL WEEVIL

ERE is a tiny, foreign creature not as big as a pea which levies a tribute of two hundred million dollars a year upon the American people, invades American soil, and collects—an insect that has changed the trade currents of the world.

Having an instinctive prejudice against paying tribute the American people has resented bearing this unapproved burden, and has resisted it. Long ago it declared war upon the invader, rallied the forces of the government and the people, summoned science to its aid, and met the enemy in the open to give battle.

For a generation the fight has gone on. Each year during that generation the invader has advanced his lines, has occupied more American territory, has levied a bigger tribute.

Today as dominant a person as Uncle Sam, most powerful person in all the world, he who cut continents apart at Panama, who bridged the Atlantic and sent his millions to fight in the World War, admits that he cannot repel this one small enemy, an insect. It must remain within his borders, he reluctantly acknowledges, and he will pay this annual tribute, but will bend his every effort to keep that tribute from growing larger—to reduce it **if**

214

possible. Yet he has no hope of avoiding this handsome payment now and indefinitely into the future.

This collector of tribute is none other than that enemy of the South, the boll weevil. Back in the nineties it entered the United States down there where Texas touches Mexico near the mouth of the Rio Grande. It started its march inland, dug itself in, each year advanced its lines from 40 to 160 miles. If the authorities had taken the advice of the entomologists in the beginning, if they had established a zone bare of cotton ahead of the weevil, it could have been stopped. But this was not done. In thirty years all that part of the United States from the Gulf of Missouri and from Mexico to North Carolina, the greatest cotton producing area in all the world, was occupied.

These little drab, hard-shelled, long-snouted weevils hide themselves in trash about the cotton fields, in adjacents woods, in the Spanish moss that festoons Southern forests, keeping themselves as warm as they may through the winter. A small percentage of them live through to emerge in the spring, breed new generations, and begin a repetition of their depredations. During the summer they fly about the cotton fields, laying their eggs in the young cotton bolls that have just begun to take form and from which the white fiber would grow later, there to hatch into grubs which will eat out their hearts, blight, and kill them. So is the great staple crop of the South injured, in some places almost exterminated.

The gipsy moth which attacks the trees of New England came from Europe. The cottony cushion scale which at one time threatened the orange crop of California was introduced from Australia. The grasshoppers which sometimes swarm over the wheat lands of the Mississippi Valley spring from the waste lands of our own Rocky Mountain regions. This invader of the cotton fields of the South is an uninvited guest originating in the far interior of Mexico.

Though the most insidious enemy which cotton has in all the world, this weevil is itself a child of cotton. Were it not for cotton every boll weevil in the world would this year die and there would be no more of its kind. When an insect is so dependent upon a single plant, that very fact is evidence that the two have lived together for many thousands of years.

It is not generally known that cotton originated in tropic America, that, as a wild plant, it first grew along that narrowed strip from Mexico to Peru where the Americas are linked together. This wild cotton, mother of the plants that we now cultivate, grew on trees. The cotton plant in its natural state is, in fact, a tree. It is at home in the tropics where there are no frosts to nip it in the autumn and so it grows throughout the year as do the trees of our orchards. When domesticated, however, and brought to a more northern clime, it is killed each year with the coming of frost and must be planted again in the spring. Thus is cultivated cotton one of man's odd developments of a plant which in nature was a thing quite different.

It is in these cotton trees of tropic America that this boll weevil has lived through the centuries until the habit has become so deep rooted that it can raise its young no place but in these cotton squares. When it crossed the Rio Grande, therefore, and found itself in the presence of innumerable cultivated cotton plants, it had found a condition more favorable to its racial development than could have been conceived by even the imaginative Jules Vernes of its race. Here was half a great nation given over to the careful cultivation of unlimited quantities of the very food upon which its existence and multiplication depended. There was latent within it the possibility of rapidly increasing its numbers. All it needed was a sufficient supply of cotton bolls in which to lay its eggs. Busily it went to work in this insect promised-land and soon there were more boll weevils in the single state of Texas than had ever before existed in all the world, and because there were boll weevils in Texas, the cotton output of that great State tragically declined, and the profits in its cultivation almost ceased to exist.

The boll weevil is a member of that great family, the beetles, those hard-shelled creatures which the scientists call *coleoptera*, which means "sheath wings." The weevils are little and inconspicuous beetles, from one-eighth to one-quarter of an inch in length, their outstanding distinction being the possession of a long snout. They go about drilling holes with these snouts, laying their eggs in the holes, and using their snouts again to tamp them in. It is such a weevil, but a little different from the boll weevil, that lays its egg in the cherry or the plum and produces the worm so often encountered in those fruits. The weevils are unusually disgusting nuisances.

When the weevil egg is laid in the cotton square, there

emerges from it in three days a tiny grub which sets about feeding on the vitals of this fountain source of cotton. The result is that the square is killed, that it eventually drops off, that it produces no cotton. Where the weevils are sufficiently abundant, entire fields produce no cotton.

The campaigns of the government for the past generation, directed by its men of science, have had as their object preventing this little insect from laying its egg in this cotton square. They have nowhere succeeded. They have held the weevil in bounds, but they have not

BOLL WEEVIL LARVA IN COTTON BOLL.

been able to drive it out. They have been able to point out to the farmer the wisest course under the circumstances, to urge him to clean up his fields in the fall so that the boll weevil may have no place in which to keep alive, to plant his cotton early so that the crop may be harvested before the weevils became numerous and, finally, to resort to a scientific method of poisoning the insects.

In the beginning there was the entomologist's hope that parasites might be found that would successfully combat the boll weevil. As a matter of fact, there are scores of parasites that do much toward maintaining the balance of nature—toward holding this insect in check. There are parasites that lay their eggs within the cotton bolls already occupied by the weevils and these eggs hatch and devour the developing grubs. In some fields it has been shown that 50 per cent. of the weevil grubs are killed by these parasites. There are other insects also which, after they are fully grown, prey upon the boll weevil grubs, regarding them as choice items of food. The most important of these are certain minute brown and yellowish ants that frequently occur in the cotton fields of the South, and may be seen running busily up the stalks, peeping inquiringly into the bolls and, where these are found to be occupied by weevil grubs, plunging in and pulling those tiny creatures out to bear them away to the ant burrow. Yet other ants attack the cotton squares that have fallen to the ground and likewise despatch their occupants. These insect enemies, however, are not sufficient to vanquish the boll weevil, and it holds its place despite them.

After a generation of experiment, the government and the cotton growers have come to the conclusion that the only method of combatting the weevil which gives assured results is the use of poison. Many years were required in working out the detailed method of getting a dose of poison to the millions of weevils that inhabit a given cotton field. Finally a scheme was perfected which is today in operation throughout the cotton region, and the employment of which is becoming steadily more nearly universal. The poison used is calcium arsenate, and the form in which it is employed is that of a finely ground powder. Machines have been developed for dusting this powder over the plants of the cotton field. The principle is the same as that of the little "gun" which is used for dusting insect powder behind the molding boards of one's own households. It is blown among the plants and there sticks and perchance finds its victim. The simplest of these machines may be borne by a laborer and operated by hand. A more complicated sort is operated between two rows of cotton by a single horse and dusts the plants on both sides of it. A yet more complicated machine straddles a row and dusts a wider swath. The highest development of the poison-dusting idea is its use by airplanes which fly low over the cotton, blowing out clouds of the powder and thus carrying the possibility of death to wide swaths of cotton fields covered.

This poison powder obviously will stick to the plants better when they are covered with dew. It is, therefore, wise to do the dusting at night or in the early morning. The weevil may eat the poison in the course of getting its breakfast or, what is more likely, may drink of the dew which has dissolved some of it and thereby meet its death.

For five dollars, it is estimated, a farmer may dust an acre of cotton four times. This number of dustings in a season properly administered is a satisfactory insurance against material injury from the boll weevil. The investment of this five dollars on an acre on the general average will return to the farmer five times that amount of money. So is the wisdom of the investment assured. So, because of this insect, does it seem that the cotton grower of the South, if he is to raise cotton at all, must accept the burden of five dollars on an acre of his crop year after year indefinitely.

This long-continued fight against the boll weevil is again an illustration of the desperate and unceasing warfare between man and the insect world—the longest, most expensive, most difficult warfare that has ever been carried on since time began. The battles in this warfare may be seen whichever way one may turn. Without shifting one's attention from the beetle hordes of which this boll weevil is a member, numberless conflicts may be seen.

There is, for example, the familiar potato bug which is in reality the potato beetle, a cousin of the boll weevil, a brilliantly striped marauder which appears on the plants and devours them. The potato bug spends the winter under ground, emerges about the time the potato plants begin to show, lays its eggs in their leaves, and these hatch into livid red larvæ for a new generation. There are two or three broods a season, and whoever grows potatoes is likely to have to wage upon them a campaign of repeated poisoning. This campaign is comparatively simple, since it consists merely of sprinkling the potato plants with powdered poison, usually Paris green.

He who attempts to grow asparagus encounters another beetle, this time an introduction from Europe, one-fourth of an inch long, black with red and yellow markings. It lives in the ground through the winter, but comes out early enough to lay its eggs in the first asparagus shoots. Slimy, greenish slugs emerge from them and eat into the plants. The asparagus grower cannot apply poison because his customers eat these tips. He must, therefore, resort to strategy. When he cuts the asparagus from his beds, he should allow a few stalks to stand. The beetles, having no other place, will lay their eggs in great numbers in these shoots. After a week they are cut and destroyed. By always leaving these few stalks as traps for beetle eggs the numbers of the pests can be kept down. But the fight must never cease.

Blister beetles are well known and worked as potato bugs before the arrival of the immigrants from Europe that are now more numerous in the gardens. These blister beetles get their name because of the fact that, if you catch one, it will throw out a liquid from its joints that will blister your fingers. This is its method of defense. This insect performs, in the course of its life cycle, some most remarkable feats. Its mother lays her eggs in some flower, a rose, for instance. They hatch into active little creatures known as triungulins. This triungulin is an acrobat and a bareback rider. He lies in wait in his rose until just the mount he wants comes along. This mount is a fuzzy bee that has its burrow in the ground. The triungulin springs lightly to the back of this bee. It does this solely for the purpose of stealing a ride. It clings on until the bee has reached home. Then, showing no gratitude for the transportation furnished, this vicious little creature alights, crowds its way into a cell which the bee has arranged for its young, eats her larva there and feasts for growing days on the food that has been provided for that larva. Eventually a beetle comes from the cell that was intended to produce a bee.

QUESTIONS

1. (a) Many of our insect pests have come into the United States from foreign countries. Illustrate.
   (b) Name the original home of the cotton boll weevil. In what section of the United States do we find it today? What physical conditions promote its spread in these regions?
   (c) Would it have been a wise thing for the cotton growers to have taken the advice of the entomologists?

2. (a) The cotton boll weevil is absolutely dependent upon cotton for its existence. What does this prove?

   (b) How could this pest be exterminated once and for all? Are the Southern planters willing? Would you be willing?

   (c) Compare this cure with the treatment of herds of fine cattle if the hoof-and-mouth disease breaks out.

3. (a) Some of our most troublesome pests are found among the weevils. Illustrate. Mention the important characteristics of this family.

   (b) Insect enemies have been found in other cases to maintain the balance of nature by preying upon the wrong-doers. Has the cotton boll weevil any enemies? Why has this source of relief been unsatisfactory in coping with the situation?

4. (a) What means has the government found to do most good in destroying this pest?

   (b) Describe the various methods used by the Southern farmer in protecting his cotton crop.

5. If you were a farmer how would you combat such problems as the potato bug, the asparagus beetle, and the blister beetle? If you did not know what to do in such a case, where would you go for advice?

6. Would it "pay" for all citizens to have a knowledge of the basic and fundamental facts of science pertaining to our common, everyday life? Prove your statement.

CHAPTER XIX
THE FLEA

TAKING it all in all, you will find that the flea, in its manner of life, has shown the rarest judgment of any of these creatures of the insect world.

It has chosen a home deep in the fur jungles of the animals on which it lives. It is a snug home entirely sheltered from rain and storm, a home not reached by the blistering sun of summer nor the rigors of winter's cold. Other insects may confine their activities to the summer months, may run their courses and die by autumn, leaving only a few to carry the race over to another spring by means of a long sleep. Not so, the flea. Its home is equally good summer or winter. It can continue its life cycles without seasonal interruption. Aside from its fur protection, its lodging is supplied with artificial heat. The heat is furnished by the landlord free of charge. The furnace never goes out. An even temperature is maintained day and night. There are no fires to build, no ashes to carry away.

In choosing its place of residence, the flea also has solved the question of keeping always on hand an abundance of food. This animal on which it lives, this model landlord of the even-heating furnace, also maintains an inexhaustible supply of the choicest of food, a food so excellent that nothing else in the world would tempt the flea to dine out. This food is none other than good, red blood. There are inexhaustible supplies of it just beneath the floor of the flea lodgings. The flea, incidentally, has a first class auger. Whenever it would dine it has only to bore a hole in the floor and drink its fill. There is never a shortage of supply—no reason for a practice of food economy. And the grease never hardens on the gravy. The food supply is kept in a thermos bottle that maintains it always at a degree of heat that accords with the dictates of flea appetites. This, in fact, is an almost faultless home which this wise insect has chosen.

DOG FLEA.

The flea does not even have to worry about transportation. These hosts run merrily about. Any time the flea wants a bit of a change, to sleep for a while in a pile of shavings, to investigate the habits, customs, and hospitality of these men creatures, it has only to hop off its mount, have its fling, and later climb back on Fido while he is asleep there on the door mat.

The flea, thus accustomed to all modern conveniences in its home, promptly moves out when there is any interruption of satisfactory service. The cat upon which it is living and feeding may die. The furnace dies down. The food gets cold. The fleas immediately move out. There are no fleas on dead cats.

Once upon a time, a hundred thousand or maybe a

million years ago, this flea had wings. There are small nubs on its sides that show where they once grew. That was before it settled down to life on hairy animals. It had to hustle in those days and needed wings to get about. But not any more. It does not now need these air agitators in its business, so it has discarded them.

The flea is a sort of distant relative of the fly. It is sometimes styled a highly specialized fly. Go back far enough and it has a common ancestor with the fly. But the flea seceded from the fly family and started an order and a civilization of its own—a race of dwellers in these moving jungles. Some entomologists still put the flea in the fly order, the *Diptera*, the "two-wings." Other scientists, however, give it the distinction of an order all by itself, and a name that is as awkward and difficult as any in the dictionary. They call it *Siphonaptera*. Anybody knows that "siphon" means tube. It is here used because the flea drinks through a straw. It is harder to guess, however, that "aptera" means "wingless," which it does, and the flea is obviously without wings. So *Siphonaptera* means "tubewingless." The flea is, therefore, Mr. Tubewingless, which, to be sure, sounds a bit awkward, but is, none the less, its scientific name.

This flea more or less deserves to be placed in an order all by itself because of its peculiar make up. Nearly all the insects, for instance, are flattened upon the surface upon which they live. Note, for example, the cockroach, the ladybug, or the butterfly. The flea, on the contrary, is flattened the other way. It is shaped something like a hatchet, but a hatchet that stands on its head with the edge forward.

There is a reason for this. The flea lives always in the jungle. If it moves at all it must split that jungle. It could not get through the hair forest at all if it was flattened like a cockroach. It could not even find a place big enough to sit down. It must have a sharpened prow like a ship.

But the jungle in which it lives is so dense as to require other aids for its penetration. The flea might crowd itself deep into the underbrush and then, when it stops pushing for a moment, its sides being slick, it would flip back to where it started. What it wants to do is to get ahead in the world. When it has made a certain advance it does not want to slip back again. To guarantee against this, therefore, it has taken thought, which might not be a bad idea for certain creatures on a much higher plane of existence. It has grown some spines on an otherwise naked body. They lie down on the body, points backward. When it has wriggled past given objects in the jungle, these spines take hold and keep it from slipping back.

These spines also serve a good purpose in helping it escape when captured. If a man happens to catch a flea between his thumb and forefinger, he is likely to notice that it is quite hard to hold. It kicks and squirms and these spines take hold and work it always to the front. Presently it is gone. Not many people know fleas well enough to advise as to details of their capture. One point might be borne in mind, however. If the fingers are kept wet while flea hunting, the chances of holding the victim will be better. If captured fleas are plunged into a bowl of water, also there is less chance of their escaping.

Another odd thing about fleas is that they are practically blind. Some of them, such as the bat flea, for instance, have no eyes at all. The dog and cat fleas have only simple eyes with which they can see not more than an inch away. It is all they need in the jungle. None of them have the compound eyes common to most insects. Their easy lives call for no long range vision. If a flea were called upon to play center field, to knit, or go to the movies, it would need better eyes. But in the indolent luxury of its well-ordered home it has needed eyes so little that they have steadily dimmed and almost disappeared.

The antennæ of the flea, as is the case with many insects, are its chief dependence in getting along. As feelers they are very handy in the jungle. Their sense of smell guides them to the animals upon which they live. They are peculiar antennæ, different from those of any other insect in that they have grooves in which they can be laid away. They are on the sides of the flea's head and, when he starts through the jungle, they would cause a lot of trouble if they stuck out as do those of most insects. But they do not. They fold back into a groove in the side of the head kept for the purpose.

There are three main families of fleas of which the human fleas of Europe should probably be named first. It makes its habitation in the houses of the people and feeds largely upon human blood. Strangely it has never secured much of a foothold in the United States, although better established in Canada. It has been suggested that Canada is more closely tied to Europe than is the United States.

Then there is the dog and cat flea which is very much

like it. There is also the bat flea and fleas that live on pigeons and poultry. These are all much alike. The dog and cat flea is much inclined to leave those animals and busy itself upon occasion with the master. The dog and cat flea is the species most often encountered in American households.

Fleas do not live on as many kinds of animals as is generally supposed. Members of the dog family, cats, rabbits, bats, pigeons, and poultry about run the gamut. Cattle, sheep, horses, and such are immune. Current opinion is to the effect that monkeys are well supplied with fleas. At the Zoo crowds watch them ministering to each other and think that they are picking fleas. This is not true at all for they have no fleas. Whatever else their purpose, they are succeeding in fooling the human throng.

The most hurtful of fleas, the Jigger, encountered in the West Indies, South America and, in fact, in some of the extreme Southern States, is unknown to most citizens of this country. It is a small flea, and the female attacks the bare feet of human beings, bores herself into them between the toes and under the toe nails. There she swells with ripening eggs until she is as big as a pea, causing great pain and inflammation.

The female flea is not as small as the male flea, the female being more highly developed in most of the creatures of the insect world. These females do not get along very well together as can be demonstrated by putting them into a test-tube together. They immediately go at each other hammer and tongs, fighting desperately.

The flea is better fitted for jumping than even that

specialist, the grasshopper. All six of its legs are geared for flinging its body into the air. Small as it is, it can jump a foot high. It can be broken of jumping, however, as has been proved by certain individuals who have taken great pains in training fleas to perform tricks. By confining them in small glass boxes where they will bump their heads every time they hop, they may be discouraged and soon broken of the hopping habit. Thereafter they will plod quietly along on their six legs.

In America the dog and cat flea are oftenest encountered in the house. They live on these pets and may breed in great number. They lay their tiny eggs in the hair of the animals and these fall off on the floor, get into the

A DOG FLEA LARVA.

carpet or into cracks or, more numerously, into the bedding of the pet, and there hatch into a tiny worm-like larva. This baby flea lives on decaying animal or vegetable matter for some days, spins itself a cocoon, and becomes a pupa. When it emerges, it is a flea, ready to find itself a model home on the first cat or dog that comes near.

Insect powder is the greatest enemy of the flea. When the family cat or dog gets them in great numbers it is necessary only to turn them on their backs and rub the insect powder into their fur to make the flea tenants most unhappy. Then the powder may be sprinkled freely about the sleeping places of these pets, even in the beds of the human occupants of the house. The fleas breathe this powder into their ventilating systems, and the result is disastrous.

Another sure method of getting rid of fleas is to trap

them. The best of traps is a dog. A dog can go about the house and, taking a quiet nap here and there, pick up a high percentage of fleas as it goes. If the dog is then washed with carbolic soap, all these fleas are killed. The dog starts out again, clean and uninhabited, and, if there are plentiful fleas about, soon returns alive with them. If he is washed again with tar soap, another instalment of the pests is killed; he may then be sent out to trap another lot. Each time he returns with fewer fleas because the supply is being lessened. If this program is continued for a few weeks, washing the dog every two days, the stock of fleas will soon be exhausted. Eventually the last fleas will be caught.

A DOG FLEA PUPA.

The flea is generally considered a creature which, in this world, plays the part of the comedian. His contact with man is usually of a burlesque nature, and man's conduct under that contact has often been used for purposes of provoking laughter.

As a matter of fact, the flea in his major rôle is a heavy tragedian, an instrument of death and destruction. His relationship to the fly family, remote as it is, is again indicated by the fact that he is a carrier of disease and as such exerts a malignant influence on man, not unlike that of the fly and the mosquito.

It is the flea that transmits that dread disease of the Middle Ages, the bubonic plague, which, though less uncontrolled today than of old, still appears now and again and takes its toll. Bubonic plague is a disease of sudden death, which may smite a victim on the street as though he were shot by a gun and cause his death after a few convulsive gulps. Its native home is India and reserve supplies of it are supposed to exist in the wild country about the Himalayas. They keep alive among the black rats which are native of that region. Black rats, however, become ship rats, and, as vessels of commerce go about the world, the pests are carried to many ports.

Bubonic plague may be prevalent among these rats. Those animals that have it would get well or die and the disease would disappear were it not for the fact that there is an insect carrier which transmits the germ of it from one rat to another and from rats to human beings. This insect carrier is the flea.

Calcutta has long been the port from which bubonic plague is broadcast. It found its way to Venice in 1403, and it was during the ravages of that time that the first quarantine was established and the basis laid for that system of quarantines which has become so much a part of the sanitary work of the world. Throughout the sixteenth century it was a permanent disease on the European continent. In 1665 it killed seventy thousand people in London alone. As recently as the first decade of the present century it appeared in the United States and got a hold that long baffled the public health authorities.

It came to San Francisco, and the method of its coming was, of course, on rats having fleas riding in ships. Realizing the danger of its being introduced in this way, many precautions were taken to prevent the landing of these foreign rats. It is known, for instance, that their customary method of coming ashore is by walking the tight-rope which ties the ship to the dock. About such ropes there are clamped obstructions which the rat cannot pass. So is an attempt made to prevent a landing. If, however, a single diseased rat infected by fleas succeeds in getting ashore, the plague may be started. This rat, being already ill, may soon die, and its parasites, in accordance with their custom, will immediately leave its body and seek new homes on living animals. So will they carry the plague to these other animals. So did the rats of San Francisco become infected. The infection spread to the prairie dogs in the back country of California and there lived in their burrows for years, and probably still lives, as it survives among the black rats of the Himalayas. Only through vigorous and unceasing campaigns were the public health authorities able to reduce it to a minimum.

Thus does a knowledge of the life story of the sprightly flea reveal the fact that here is another insect demon frightfully dangerous to the well-being of man, that here is another evidence of the correctness of the theory that insects are man's worst enemy.

QUESTIONS

1. The author cites many illustrations which would make us conclude that the flea lives in a model home. Do you agree with the author? How has the flea solved the food, shelter, fuel, and transportation problems?
2. Did the flea always enjoy the conveniences of life and participate in the luxuries of the animal world? Account for some scientists classifying the flea under Diptera while others put the flea in an order by itself.
3. (a) Tell how the flea is fitted for life in a hair forest or jungle such as Fido's back resembles.
   (b) Describe the flea's eyes, its antennæ, the purpose its spines serve, how the flea is fitted for jumping.

4. List the three main families of fleas—the habitation and countries in which they are found. What animals are immune from fleas?

5. Suppose your dog suffered from fleas. How would you go about relieving him and ridding the place of fleas?

6. (a) India is the native home of the flea. Explain the rôle of the flea in transmitting that deadly disease known as bubonic plague. What brought about the first quarantine established in Venice (1403)?

   (b) How have the public health authorities in the United States, particularly in San Francisco, been baffled by this problem of the flea? What precautions have been taken to prevent black rats from entering our ports and thus increasing ravages from the plague?

   (c) Having learned that insects are among man's worst enemies, can you give another reason for observing "Clean Up Week" every week in the year?

Chapter XX

THE JAPANESE BEETLE

UT the point of your compass at Riverton, there in New Jersey across the Delaware from Philadelphia, and draw a circle with a twenty-five mile radius and you will have outlined an area in the United States containing a more deadly menace than if it were underlaid with TNT and the fuse were burning short.

Within this circle and existing no place else on the Western Hemisphere is an insect plague, the Japanese beetle, now indestructibly dug in and prepared for an annual widening of the circle which cannot be stopped by force, cunning, or scientific understanding at present in existence, but which seems destined to go steadily forward creating a devastation more deadly than that of any Sherman marching to the sea.

Within this circle a few inches below the surface of the ground in the late spring there will be found a layer of greedy grubs gnawing at the grass roots. In many places they are so thick that 1500 of them may be found in a measured square yard of sod. When thick enough they cut the roots of this sod below the surface as cleverly as it might be done with a cleaver, so completely that it may be rolled up as a blanket on your bed. Thus they blight the lawns and golf courses in occupied territory and convert them into sere wastes.

These grubs, having eaten their fill, change themselves into chrysales, as insects do, to go to sleep for a few weeks and then awake in a different form—in the form of brilliant, green beetles with dull brown backs. These are the grubs grown great, the above ground dwelling adults that develop from the dirt-bound young. These are the members of the marching forces which will carry the invasion each year into new territory, there to establish new generations to dwell perpetually.

When the grub has turned to a beetle, this creature has, for that year, completed its work of devastation at the grass roots. This green beetle soon takes flight to the tops of the trees that border adjacent roads, to nearby orchards and vineyards, to corn and clover fields of the vicinity, to the shrubbery of the gardens that surround the homes of the inhabitants. On this sort of vegetation they congregate in countless thousands. They are hungry creatures and immediately set to work eating the leaves of these plants. They devour the main body of these leaves and there remains only the delicate lacework of the ribs that support them. The leaves are the lungs of trees, and without them they cannot live. So where the beetle attack quite covers a tree, it dies.

PUPA OF JAPANESE BEETLE.

A little later, by the time these beetles become particularly abundant, the fruit of the orchards and the ear of the sweet corn will be ready to provide more substantial clusters of nourishment. To these the beetles will resort and, where the infestation is well established, an organized fight will be necessary to prevent them from getting the whole crop. Such is the menace faced in the Philadelphia district.

It all began back in 1916 when it was discovered that a little area of half a square mile about the establishment of one nurseryman was infected by these beetles. The story of the first period of the advance of the plague may be written in tabular form, showing the progressive increase of the area occupied during the first eight years of its presence in America. It would appear like this:

$$
\begin{array}{ll}
1916. \ldots \ldots \ldots \frac{1}{2} \text{ square mile} \\
1917. \ldots \ldots \ldots 3 \text{ square miles} \\
1918. \ldots \ldots \ldots 6 \text{ square miles} \\
1919. \ldots \ldots \ldots 48 \text{ square miles} \\
1920. \ldots \ldots \ldots 100 \text{ square miles} \\
1921. \ldots \ldots \ldots 260 \text{ square miles} \\
1922. \ldots \ldots \ldots 770 \text{ square miles} \\
1923. \ldots \ldots \ldots 2500 \text{ square miles}
\end{array}
$$

Early in the summers of subsequent years the beetles went over the top for the annexation of additional territory. The marshalled forces of the State of Pennsylvania, the State of New Jersey, and the Federal government regularly gave battle. They admitted in advance, however, that despite their best efforts they expected the beetle each year to annex a territory as great as that which it already occupied. From the start there was alarm at the apparent probability that each year would witness an advance and that a time would come when this creature would crowd far into the agricultural heart of the nation.

This dramatic situation came about because of a very

small incident, because of the fact that a single nursery-man brought to the United States certain plants of the Japanese iris in which slumbered a few of the young of this beetle. The iris were planted at Riverton; the grubs

AN ELM LEAF ATTACKED BY JAPANESE BEETLE.

developed into beetles, and this started the widening circle which has grown year by year.

Science has sought in various ways to check this beetle. It has tried to find a poison so economical that it might be possible to spread it upon the affected lawns and meadows and thus kill the grubs before they become beetles.

It has, in fact, prepared an emulsion of carbon bisulphide which may be thus used and which gets satisfactory results. A treatment of grass areas with this emulsion is, however, so expensive that it is impracticable except on small lawns and such areas as the putting greens of golf courses.

The scientists have also sought a means of destroying the beetles after they have taken to the trees and shrubs. They have naturally resorted to spraying with insecticides since this is the demonstrated method of reaching most of the destructive insects. The Japanese beetle, however, has proved itself a canny individual and, when the spraying starts in its vicinity, it takes to its wings and flies away. It agrees to leave the trees that are sprayed, but it refuses to allow itself to be poisoned. Thus spraying is a temporary protection to trees, but not a destroyer of the pest.

An example of the abundance of these beetles was furnished one July morning in an orchard at Riverton containing 156 peach trees. It had been found that of an early morning the insects are inactive. If the trees are vigorously shaken, they will fall to the ground and may be swept up and destroyed. In two hours of this July morning in this orchard 13 tubfuls of these insects were so collected, each tub holding 16 gallons. Much to the chagrin of the orchardman, however, he found that, on the following morning, beetles were as plentiful in his trees as before.

Aside from the steady pushing back of the rim of the circle of the infested area year by year, there faces the authorities the danger that these beetles may travel great

distances and start fresh nuclei.   Much of the area that
furnishes the fresh fruit and vegetables for the city of
Philadelphia is infested.   As this food-stuff is hurried to
the markets on early summer mornings, it is unavoidable
that many beetles should not be carried with it.   These
beetles, for instance, when they attack sweet corn, dig
in beneath the shucks and there remain and feast.   When
the sweet corn comes to market, many of them arrive
with it.   About all the markets of Philadelphia and in the
streets and gutters adjacent are likely to be found great

numbers of these Japanese beetles
going about and eating the waste,
quite unconscious of the fact that
they are part of a warfare with man
that will go on for generations.
Philadelphia, being a distributing
center, thus offers the possibility of
sending this menace to distant sections, there to get a foothold.   Preventing this possibility was one of

AN APPLE ATTACKED BY
JAPANESE BEETLE.

the serious tasks facing the State and National authorities.

The law of the insect world, that the greatest enemies
of insects are other insects, held, the scientists figured, in
the case of the Japanese beetles.   In Japan these beetles
do little harm, which fact is probably due to the presence
in Japan of some other insect which preys upon them and
keeps them within bounds.   When, however, an insect
is introduced into a region where it has not previously
existed, it temporarily escapes its parasites and, therefore, breeds much more rapidly than at home.   The

Japanese beetles in America are enjoying this immunity from attack by their customary enemies.

The United States Government, upon discovering the presence of the plague, hurried a corps of experts to Japan, where, in coöperation with Japanese scientists, they pried into the everyday life of this beetle to find where and by what it is attacked. The results were that they found certain parasitic flies and wasps in Japan that set upon these beetles and destroyed them. The usual point of attack was upon the beetle when in the grub stage. These tiny flies and wasps, knowing by instinct that the beetle grub is beneath the surface of a grassy meadow, fly about that meadow and lay their eggs. These eggs hatch into much tinier grubs of these flies and wasps, which immediately dig into the ground, find the beetle grub, bore into its body, use it as entrée, salad, and dessert until it is done to death.

The government acted most energetically under the circumstances. Soon there were in course of transit from Japan to the Philadelphia area shipments of many tons of earth which was inhabited by beetle grubs, which, in turn, were inhabited by these parasites. These tons of earth rode on the fastest steamers, and the fastest trains, that they might reach the infested area before these parasite grubs turned to parasite flies and needed the open spaces for the establishment of new generations of their kind.

It is upon these parasites that the experts base their chief hope. It will take years for them to become established and so multiply as to become effective, just as it took years for the parasites of the gipsy moth to turn the tide of the war upon that pest. If, however, the parasites prove not to be effective and the plague goes on and on unchecked, American communities will be forced to the necessity of getting on as best they may in the permanent presence of this menacing insect, just as they raise cotton in the presence of that other beetle, the boll weevil, and just as they raise potatoes in the presence of another beetle, the potato bug.

QUESTIONS

1. (a) How did the Japanese beetle gain access into the United States?
   (b) What do scientists know about this destructive beetle? Consider the three stages of this insect—grub, chrysalis, beetle.
2. Explain how lawns and golf courses are devastated by the insect when in the "grub stage."
3. (a) What damage is brought about by the visit of beetles in orchards and vineyards?
   (b) Tell of the experience of an orchardman in Riverton who endeavored to rid his peach trees of Japanese beetles.
4. Study the table showing the advance of the plague in eight years. Of what significance to science is this table?
5. What remedies have scientists given us for this pest? Can they be used, and do they get results?
6. Why are Philadelphia and other places adjacent to Riverton, New Jersey, in grave danger of being attacked by this menacing insect?
7. (a) How are experts trying to bring about a "balance of nature" in this problem?
   (b) On what do they base their chief hope?

CHAPTER XXI

THE PEACH MOTH

HE United States recently has been successfully invaded by an enemy from abroad, which has captured Washington, its capital city, together with the surrounding coun· try for fifty miles in every direction, has defiantly dug itself in, and is making faces at this, the most powerful nation in all the world.

The invader is an insect, the peach moth, which each year launches its campaign of destruction and blights this popular crop. The best authorities of the government, having studied it carefully, have come to the realization that it may not be possible to stop it, that it will steadily spread, that it will overrun all the peach-growing area of the Atlantic States, that a time may soon come, so deadly are these insect enemies, when peaches, because of it, will disappear from American tables.

The government has declared war on the peach moth. Its declaration is no small matter. It means the beginning of another of those warfares between an established government and members of the insect world which are not uncommon, that often go on for decades, that sometimes result in the defeat of great nations. The unchecked invasion of the boll weevil in the South, fought at every foot across a continent, and the gipsy moth in

New England, are examples of the inability of the great American Government to repell these attacks.

This peach moth entered the United States surreptitiously despite the fact that the government forces were watching every border. In doing so it marred what was otherwise a happy interchange of courtesies between the United States and Japan.

It will be remembered that, back in the days when genial Mr. Taft was President, none other than the Empress of Japan presented Mrs. Taft with a selection of flowering cherry trees, the choicest of those ornamentals that have come to typify that oriental nation. When they reached the United States, those watchful authorities who try to prevent the introduction of insects that might be harmful examined them and found them to be so infested with a number of insects that it was thought advisable to destroy them. The Empress, no whit discouraged, later sent over a much larger shipment of these same cherry trees. Great care was taken in Japan to make sure that they were not infected. When they reached the United States, they were again inspected. In addition, they were thoroughly fumigated, given that treatment which was then considered absolute insurance against any creature coming through it alive.

The cherry trees were brought to Washington, were set out along the Speedway and around the Tidal Basin, in that parkway beyond the Washington Monument and the Lincoln Memorial where the city looks across the historic Potomac to Virginia. They have since come into such magnificent maturity and beauty that tourists come in the spring by thousands to see them, and pictures of them are the favorite illustrations for Sunday supplements.

Three or four years after these trees were planted a scientist in the Department of Agriculture found in its grounds, half a mile from the Speedway, a strange moth. These scientists know that a moth is a newcomer just as the average citizen would recognize a wandering Hottentot as a sort of man with whom he was unfamiliar. A strange insect always has in it possibilities of doing great harm, so this one was led into the laboratory and given the third degree.

In the meantime the country roundabout was dragged for others of its kind and more were found down among the cherries, and a few a mile away across the river where the government maintains Arlington Farm for purposes of agricultural experiment.

The suspect was carefully examined for identification. There was nobody in the United States, it developed, who could place the stranger. European scientists were called in, but with no greater success. Finally specimens were sent to the other side of the world, to Japan, and there others were found just like it. It was a Japanese species, the Americans said. The Japanese admitted that it was present in Japan, but disclaimed ownership, stating that it had been brought to their island from Australia. It seemed, upon investigation, that it is a native of Australia.

The peach moth has been convicted upon circumstantial evidence of gaining entry on cherry trees. Like all the members of the moth family it first appears in the world as an egg. This egg becomes a larva which devotes itself chiefly to the business of eating and growing. When it has become a fat sac of tree juice, it crawls into a crack in the bark and there weaves a cocoon for itself in which it goes to sleep. As it sleeps, the substance that is within it is subjected to a miracle of nature and is transformed into something near to a fairy, into a glistening and beautiful moth.

The peach moths smuggled themselves into the United States while they were in their cocoons, in the chrysalis stage. While thus sleeping they breathe not at all or hardly at all. This makes it hard to kill them by fumigation. In addition to this their cocoons are water-tight

PEACH MOTH LARVA.

and practically air-tight. They undoubtedly got in while their inactive state protected them from the dangers of fumigation.

By 1915 the peach moths were plentiful at Arlington Farm. In another year they were being sent in from all about Washington by frantic fruit growers who wanted to know what this creature was that was ruining their orchards. Each year the circle about Washington widened.

It is probable that other introductions were made at about this time, for the infection soon was found in the nursery and orchard section of Delaware and New Jersey. A touch of it was found on Long Island where there was a nurseryman with a record of having imported cherry trees from Japan. It was reported from Memphis, Tennessee, and finally came to light at Valdosta, Georgia, not 50 miles from the greatest peach belt on this hemisphere.

The air-tight cocoons of the peach moth hatch out in the spring, and the moth starts about laying its eggs on the twigs of trees. The egg hatches, and the larva crawls out to the tender tip of a twig and begins eating into it. It starts down through the center of the twig, eating as it goes. It blights that twig and stops its growth. Each larva is likely to spoil three or four twig tips before it goes to sleep to wake up a moth to breed other blighting larvæ.

A PEACH INJURED BY PEACH MOTH LARVA.

Then, later, when the twigs are no longer tender, there is the fruit itself, which is good food for the grubs. They begin to bore into the peaches when they are half grown, causing them to become defective. They attack them when they are nearly ripe and penetrate to the seed. During bad years they make all the late maturing peaches in the orchard wormy. They spoil the crop. They also attack apples, pears, and cherries and are particularly fond of quinces. One often finds twenty grubs in a quince.

The peach moth is much like the codling moth that attacks apples and makes them wormy and that has long been with us, which costs the apple growers in this country $25,000,000 a year. The apple grower fights the codling-moth by spraying the trees with insecticides. It works pretty well. The same scheme was tried in fighting the peach moth. The peach, however, has fuzzy skin and it is impossible to coat it with poison as can be done with apples. The spraying method does not work very well with peaches.

The resort here, again, is to the natural enemies of these insects, which are other insects. There are, for instance, the parasitic four-winged flies, a group of the most fiendish little creatures in all the world. They go about laying eggs in or on the larvæ of these moths. These eggs hatch out and dig in. There may be twenty of them in one larva. They eat it alive, and, when it can no longer carry on, they eat it dead.

When a new species is introduced like this peach moth, the parasitic flies do not know about it. They have the habit of feeding on other species, so that it is likely to remain undisturbed. For this reason a newly introduced pest is likely to find itself, temporarily at least, at a great advantage. The customary procedure of the government is to hurry to the country from which the pest came, find out what was its parasite enemy there, and introduce these parasites. It takes the parasites years to get established, however, and there is nothing to do but to wait to get the result.

In the meantime it is observed, happily, that some of the executioner parasites that are the outstanding enemies of the codling-moth are turning their attention to this peach moth. This is the first good news in the campaign. Every ditch is being contested, and the situation looks desperate. This fuzzy little moth seems about to give our Uncle Samuel quite a tussle, and it may soon be good-by to peaches and cream in America.

QUESTIONS

1. (a) You know how oranges were saved from "the jaws of death" by the application of science. What dreadful war is being waged upon the peach moth?
    (b) Do you think peaches will be saved for Americans? Give reasons for or against.
2. (a) How did the peach moths gain free transportation into the United States?
    (b) Explain how they escaped detection and death at the hands of our experts.
3. (a) What does the writer mean by saying "the little stranger moth was taken into the scientific laboratory and given the third degree"?
    (b) Describe the life history of this moth. Which stages are most dangerous to the fruit grower?
4. List all the fruits threatened by the peach moth.
5. The enemies of this insect have been imported from Japan, but as yet they are not functioning. Explain this situation.

CHAPTER XXII

BUGS

"**B**LACKE bugges" of the night used in olden English times to be terrifying creatures of the imagination—hobgoblins, bugbears. Old-fashioned people who have clung to and handed down a family legend sort of information still frighten their children with boogers and boogermen. Even in Shakespeare's time a bug was not an insect, but a vague, weird, fearsome, unseen, hideous thing of the dark.

About this time British ships began to bring spices from the Orient and sugar from the hot countries. With these cargoes came new insects—weird and ugly creatures from unknown lands, likewise bugbears and hobgoblins. They had no names in English, so they were called, as were the terrors of the dark, bugs.

These creatures were mostly of a certain unpleasant sort of insects—were parasites upon man and beasts. The familiar bedbug was among them. So was the body louse, the "cootie" of the World War. These lived by sucking the blood of man. They had cousins that lived on the larger animals and other cousins that lived on plants. Among the most important of these latter were the plant lice or aphids.

The aphids suck the sap or plants. They are related to the scale insects, the cottony-cushion scale, for instance, that was destroying the orange groves of California until the ladybird beetle was introduced and ate it up. They are related to the stink bug that one encounters while berrying, and to that oddly powdered creature of the hothouse, the mealy bug.

All these insects have bills for piercing and pumps for extracting blood or sap. Most of them have wings, one-half of which are of hardened chitin like those of the beetle, and half of which are of membrane like those of the bee. Because of this wing structure they are called *Hemiptera*, which means "half wing." They are the true bugs. They are the only insects that should be referred to as bugs. All the six-legged animals are insects, but only these half-wings are bugs.

But, the objector is sure to say, these pests that bite man have no wings at all. True enough, but they once had. It is an odd fact that in nature unused parts tend to shrink away and cease to exist. These detestable creatures long ago became deadbeats, just as did the fleas. They stopped hustling for themselves and came to live on other and bigger animals. They had no longer any use for their wings, so these disappeared just as the muscles of an athlete will waste away if he fails to use them, or a good brain will become sluggish and stupid if its possessor does not keep it exercised by giving it a bit of wholesome thinking to do.

Of all the bugs, the aphids are probably the most interesting. There are, however, many quaint creatures of this order, and before telling the life story of the aphid, I would like to give you a look at a few of the others.

There are, for instance, a number of bugs that live ex-

clusively in or about the water. There is the water spider
or water skater, for example, which has the remarkable
faculty of being able to walk on the surface of the water,
its feet merely making dimples in that surface, but not
penetrating its somewhat tense surface film. On this
film they are truly skaters, and their exploits are marvelous to behold.

The water boatman and its relative, the back swimmer,
which are found in still pools, are also bugs. These are

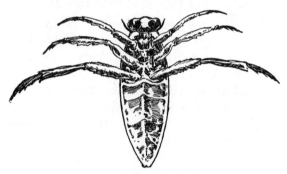

THE WATER BOATMAN.

thick creatures rarely more than half an inch long, the
legs of which are fitted and worked for all the world like the
oars of a boat. The odd thing about this back swimmer is
that he operates topsy turvy all the time making an odd
appearance with his back toward the bottom of the stream.
Both these bugs have a very odd trick of collecting bubbles
of air which they attach to the fine hairs that circle their
bodies and which they are able to take down with them
when they go to the bottom. Great numbers of these
air bubbles give them the appearance of being beauti-

fully silver trimmed. These water bugs are air breathers, but the hunting is better deeper down in the water, and, with these air bubbles as a reserve supply, they can stay at the bottom for a remarkable length of time.

In the water is to be found also the giant water bug, also a relative of the boatman, sometimes two and a half inches long and the biggest of all the bugs. Both the boatman and the giant develop wings at one stage of their lives and take to the air. It is at this time that the giant water bug becomes the electric light bug, and in the late summer he is likely to be found thumping foolishly about the glaring electric lights that may be near his home.

One of the oddest things in nature is the practice of the female of a kindred species of water bug in the peculiar manner in which she deposits her eggs. She forces the male, apparently much against his will, to sit quietly while she securely cements the scores of her eggs to his back. Thus is the father required to carry the eggs, conspicuously plastered upon his back, about with him, until the time of their hatching.

There are many odd and interesting creatures among these sap-sucking bugs which live upon plants. It will be remembered that the cicada, that curious seventeen-year locust, is a member of the *Hemiptera* and, therefore, a sap-sucking bug—the giant among plant feeders of this order. At the other extreme are the minute creatures which, in colonies of infinite number, likewise with their beaks sinking into the flowing sap, make up the bark scale.

Quaintest of all these sap suckers are the tree hoppers,

the brownies of the insect world. These tree hoppers attach themselves to various sorts of trees where they have developed resemblances to seeds, thorns, or bits of bark. They sit tight and feast to the end of their lives without arousing suspicion.

One might carefully examine a cherry tree that was full of these tree hoppers and never suspect their existence, so perfect is their imitation of parts of the tree. If one knows these tree hoppers, however, he may select wisely what appears to be a thorn, but is really an insect in disguise. He may press the mouth of a milk bottle down over that thorn and then stir it to life. He will find that he has captured in that bottle a very charming little mimic.

Since the thorns, the seeds, and the different protrusions on different sorts of trees are very different in appearance, so is there an endless variety of these tree hoppers, each imitating some part of the particular tree on which it lives. A row of different varieties of these insects in their quaint and curious forms looks like nothing so much as the creations of an imaginative artist.

Maintaining a reputation for unpleasantness are the various varieties of ill-smelling bugs—the squash bug, the stink bug of the berry patch, and many others—flat, turtle-shaped, dull-colored creatures a quarter of an inch long—and the harlequin cabbage bug, the calico back, brilliantly colored denizens of the garden. Then there is the chinch bug of the wheat and corn field which does $20,000,000 a year damage to the farmer in this country. All these insects, it should be understood, secrete this bad odor, which is also a bad taste, so as to make themselves unpopular with the birds so these latter will not select them as food. But for this bad taste they probably would have ceased to exist long ago.

Then there is the spittle insect, which is a further advance toward the aphid. Who has not seen splatches of froth on weeds, grass, and shrubs and, possibly, looked closely to try to find out what it was. Inside it, more than likely, the observer found a tiny insect a third of an inch long. The insect secretes a liquid through the walls of its body and then shakes its body until that liquid is beaten into a froth. It does not appear that this froth does the insect any particular good and its purpose is not understood.

Here appears the aphid, of which there are millions in nearly every garden, plump-bodied, pale-green bugs less than a quarter of an inch long, the green fly of the con-

AN APHID.

servatory, the "blight" of the alder, beach, or elm, the milch cow of the ants, one of the most remarkable of insects. These aphids are remarkable for a number of things, first among which is the presence of females only during the greater part of their existence. They appear first in the spring and there are no males at all. They are all just alike. Despite this fact they begin to produce new generations of aphids with great rapidity. The young are usually born not as eggs, but as active little insects,

ready to go to work. So rapidly do they mature and produce others of their kind that a garden containing a single
aphid on Monday may by Wednesday be ateem with
them. Every individual aphid of the entire crew inserts
its bill through the live skin of some tree or shrub or grass
blade, and begins drinking sap. No particular harm is
done unless the aphids become so numerous that they
drain the vitality of the plant to which they are attached
to such an extent that it is unable to serve its normal
purpose. This is very likely to happen, and does often
happen wherever the balance of nature is put out of joint.
There are, however, other insects as, for instance, the
lady-bug, which feeds upon these aphids, eating great
numbers of them. There are certain flies which lay their
eggs among them and these eggs hatch into hungry grubs
that devour them. There are certain birds, as, for example, the brilliant oriole, which like nothing better than
to alight upon a rose bush, travel about it and pick an
abundance of aphids for their breakfast.

The odd thing about these numerous generations of
female aphids is the fact that a time comes, toward the
end of the season, when the long-continued program of the
earlier months is broken and there appears, suddenly,
new sorts of creatures that have not before existed in
the tribe. The aphid generations of the autumn are both
male and female. Oddly, the generation which they produce follows a quite different scheme in its development.
This mother aphid of the autumn lays but a single egg.
That egg remains as such throughout the winter. It
hatches out in the spring into female aphids that produce,
without the presence of a male, the myriads of other fe-

male individuals that make up the mass of the aphid population through the summer.

Another odd and remarkable thing about these aphids is that during the earlier generations of the season they are wingless. They may have started their life, for instance, in a plum tree in the spring. There they may have gone on multiplying until there are great numbers of them. They have been without wings and, therefore, without the possibility of foraging for other food supplies. Within them, however, Nature has set an alarm clock which now proclaims the fact that they need to migrate to pastures new, to find new fields upon which to feed.

So, of a sudden, there appears a generation of aphids which has wings. They spread these wings and fly away, perhaps to a hop field nearby where the green and growing plants furnish an abundance of fit food. Having made their trip, the succeeding generation is again wingless, there being no occasion for it to travel. Here in this hop field they multiply greatly and perhaps spoil the crop by their numbers. Even though this does not take place, the time comes when the hops mature and no longer furnish the sap which the aphids require. Realizing the approach of this season, these creatures, still consisting of females only, breed another generation which has wings, and this generation takes flight and returns to the trees round about, and to other plants in which the sap is still flowing. Nobody knows how the warning is given of the approach of a food shortage. Nobody knows how these tiny creatures are able, at will, to cause their babies that have hitherto been wingless to suddenly develop these aids to flight. Nature knows, it seems, and is able to meet the need. Had man been able thus to add spare parts to his make-up as needed, there would have been no occasion for inventing the airplane.

These aphids secrete two kinds of materials. In the first place they produce wax with which they envelop themselves as in a cloud. They spray forth this wax. They have a carburetor through which they pass it which reduces it to minute particles. In this form it best serves their purpose. Then they secrete honey dew—which has been described as the national dish of the ants. This is a sweetened water which they extract from the sap they drink. The sidewalk under a tree inhabited by aphids may be covered by the showers of this honey which they have dropped. Ants follow these aphids about and lap up this honey. They even have flocks of them and milk them regularly.

Aphids became so numerous on fruit and other trees in gardens that it became necessary to spray them with insecticides to kill them. The hop aphid, a yellow mite one-twenty-fifth of an inch long, whose summer journey from the cherry tree and back has been described, is beyond this treatment when it once gets into the hop fields. These mites are destructive to many crops. One of them, for instance, is the plague of vineyards. They attach themselves to the roots of the vines, form tubercles, and cause their death. They invaded France shortly after the Germans had done so back in 1871 and killed 3,000,000 acres of vines. This insect invasion caused a greater loss to the French than did the German invasion, yet history has left little record of it. So have the attacks of insect enemies of man escaped especial comment.

QUESTIONS

1. (a) What is the literary significance of the expression "blacke bug-
      ges"?
   (b) Has the term "bug" used by the average person the same mean-
       ing as when employed by men of science?   Illustrate.
2. (a) Make a list of all the true bugs with which you are familiar.
   (b) Account for their wing structure.
3. (a) People often speak of bugs as biting, but this is not true.   Des-
       cribe their mouth parts.
   (b) How do bugs repel their enemies?   Compare with monarch butter-
       flies and cockroaches in this respect.
4. (a) Have you ever tried to catch a water skater?   Tell your experi-
       ence.
   (b) What makes the water boatman look as if it were clothed in a
       silver dress?   How does it breathe?
   (c) Did electric light bugs exist before electric lights came into gen-
       eral use?
   (d) How does Mrs. Giant Water Bug make Father work?
5. (a) Describe the brownies of the insect world.
   (b) What policy do the aphids seem to follow when the family larder
       gets empty?
   (c) Name all the enemies of the aphids.

Chapter XXIII

INSECTS IN GENERAL

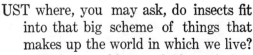

J UST where, you may ask, do insects fit into that big scheme of things that makes up the world in which we live?

If that world were laid out before us like the pattern of a crazy quilt, where would the patch which represents insects come in?

And if it came down here in the left-hand corner with the dingo dog off to the left and a cockle-bur bush south by southwest, how would we know that it is so?

In struggling toward an answer to such questions as these it may be said that we know about these riddles because scientists have studied insects, along with other forms of life, and have found that there is a plan running through all that there is in the world, that there are relationships between one thing and another and that these relationships can be shown. It is known, for instance, that the ant and the bee are cousins. There is even a relationship between the dingo and the cockle-bur bush. They are both living creatures and are, therefore, of quite a different breed from monkey wrenches and brickbats which have never felt the stir of life.

All that there is in the world is subject to classification. These classes are not so many that one may not come to know something about them. But, on the other hand, there are so many individuals that it would be impossible for the smartest man in the world to ever come to know them. For example: in the classification of animals there is a cat family. Everybody knows that cats have claws and eat flesh. One might be given a lynx for a pet and might never have heard of a lynx before. If he were told, however, that it was a member of the cat family, he would immediately know that he should feed it raw meat and beware of being scratched.

Understanding in general is largely based upon orderly knowledge, and back of such knowledge is this interesting game of classification, of putting things in their proper places, of finding their relation to other things.

If a horse is offered you for purchase, you immediately begin to classify it. Is it a draft horse or a saddle pony? Is it old or young? Is it wild or gentle? If a man is being considered for a place in the firm, or for marriage into the family, he likewise undergoes classification. Is he a white man, black man, yellow man? Is he educated or ignorant? Is he a successful man or a failure? By putting him into these classifications of known types of men, an estimate of his value can be made. Organized information can be applied to him.

So is it in attempting to measure the importance to the world of a group of life that is a part of it, a group, for instance, like that made up of the insects. For understanding there must be classification, organized information. Insects must be separated from the great mass that is the world, must be put into their own pigeonhole, so arranged that they may be compared with the other elements that join with them in making up the whole.

Starting on the big task of classifying all that there is
18

in the world that we know, all those elements that go to make it up and broadly spoken of as "matter," we find that this matter is primarily divided into two groups. One group is made up of things that have life in them, and the other is made up of those that do not have life. There is the tall pine tree, for instance, growing there on the mountainside. It has within it the spark of life; it grows, adds to itself, may suffer injury, death, decay. It is a living thing. It is animate.

Right beside it rises a rugged cliff of stone, lifeless, unchanging, incapable of feeling, of suffering injury. It is without life, inanimate.

Yonder in the field is a horse pulling a plow. But a little while ago it was a small creature, new-born, merely a promise of what it is today. As the months have passed it has added to its strength, has multiplied its size, has come to be the powerful creature it is, capable of dragging this plow down its furrow. But a little while longer and it will suffer other changes; death will come; its body will disappear, will be absorbed back into the elements. It is an animal, a live creature, capable of change, of action, of injury, of decay. It is animate.

The earth which it turns over with the plow, the earth from which grows the grass upon which it feeds, is, on the other hand, an inanimate thing, a thing without life, incapable of action, of adding to itself, of decay. It has been where it is through the lives of innumerable generations of horses. It cannot move of its own will. Unless some power outside itself takes a hand, it will lie there unchanged through unmeasured time of the future. It is without life, inanimate.

Thus does it appear that all that is in the world is subject to a division into two classes of matter, that which has life and that which has not. Here, then, is the first classification, when one begins to try to put the things about him into the groups to which they belong. A friend might wire that he is sending you a shipment of stalagmites. You might not know what they were, but if you knew to which of these broad classifications they belonged you would know whether to enlarge your zoo or your museum.

Thus does it come to pass that the first bit of a diagram may be drawn indicating this first grouping of the elements that go to make up the world in which we live. The diagram might take this form:

```
                    ┌─────────────────┐
                    │   ALL MATTER    │
                    └─────────────────┘
          ┌──────────────────┴──────────────────┐
┌─────────────────────┐              ┌─────────────────────┐
│ THAT WITHOUT LIFE   │              │  THAT WITH LIFE     │
└─────────────────────┘              └─────────────────────┘
```

The subject in which we are interested—insects—obviously belongs to that division of matter which goes to make up the part of the world which has life. Insects are animate creatures. They live, grow, die, decay.

When one gives attention to living things, a second classification immediately presents itself. A bird may sit on the limb of a tree. Both the bird and the tree are alive. Each is growing, each may be injured, each may die and decay. There is, however, an obvious difference in the sort of life which is given to the tree and that which is given to the bird. The tree remains always in the same place. It has no power of motion, may not act of its own will. The bird, on the contrary, can leave the branch on which it sits and fly away to another perch.

On the face of it, this is the primary difference between plants which are alive and animals which are alive. Animals are given the power of moving from place to place as they wish. That they may use this power they have developed certain faculties which the plants do not possess. The bird, to fly away, must be able to see that it may decide upon the point to which it will fly. It must have an intelligence to direct its course which is not necessary to the plant. In living the sort of life it does the bird has developed certain abilities. So have the other animals. The fish must swim; the antelope must run; the rat must dig; the squirrel must climb.

All such abilities distinguish animals from plants. They lead to a division of all living things into two parts, the vegetable kingdom and the animal kingdom. If one knows an object belongs to one or the other of these kingdoms he has a good deal of information about it. Thus, we have the second diagram:

Scientists have organized the study of all the life of the world and have called that study Biology. They have narrowed that study as life has been divided into its logical parts. The study of plant life has been called Botany and the study of animal life has been called Zoölogy.

Now, taking all the members of this animal kingdom and making another classification which gets still further down toward these insects which we are studying, we find that living creatures, capable of moving of their own free will, divide themselves into groups, which the zoölogists call branches or phyla (singular phylus, plural phyla), according to certain outstanding differences in the manner in which they are put together. There is, for instance, one great branch or phylus of the animal kingdom which is known as the vertebrates, the creatures which have backbones. Man has a backbone and, therefore, belongs among the vertebrates. He need not feel especially set up by the fact, however, because snakes likewise have backbones, as have fishes and birds. Another great branch of the animal kingdom is the mollusks, these soft-bodied creatures like clams, oysters and snails which are covered with shells for their protection. They are very different from the animals with backbones.

Altogether there are twelve of these branches, first divisions of the animal kingdom. Among them are the arthropods. This word means, literally, "jointed feet." The arthropods have more than jointed feet, however. They have jointed bodies and jointed legs. They are, in fact, made up entirely of segments, are put together like the cars of a train. Observe, for instance, the lobster. Here are three divisions of the animal kingdom:

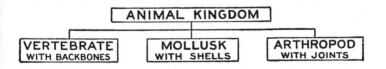

These vertebrates, those creatures having backbones, have an entirely different scheme of construction from the arthropods, the jointed animals. They have bones on the inside of them and the muscles attach themselves to those bones on the outside like ropes to pulleys, cover the bones. The bones keep the body stiff and the muscles pull parts of it about, as, for instance, the hand of a man, just as the ropes work the scoop on a steam shovel.

The mollusks, again, have an entirely different scheme of construction. They are soft-bodied and carry their houses on their backs, as, for example, the snail in the back yard. What little movement it is capable of is brought about by the expansion and contraction of bits of muscle. The shell with which it is covered serves merely the purpose of protecting its body.

But with these jointed animals, of which the lobster is a familiar example, the plan is quite different. They have a skeleton that may be compared with that of the vertebrates, but they wear it on the outside instead of on the inside. Instead of looking like the bones of the higher animals, it appears more like a crust. It serves the purpose of the bones, however, provides the stiff parts to which the muscles may be attached, forming pulleys to move the members about. But these pulleys work on the inside of these stiff segments. They are working well, however, when the cockroach goes scuttling across the kitchen floor or the rapid vibration of a bee's wings set up a hum like that of a tiny airplane.

So, as classification goes forward, one comes to know that any creature set down as an arthropod has certain qualities, jointed bodies and legs, skeleton on the outside, etc. Knowing a branch, one has much knowledge of an individual in it.

Among the jointed creatures the next systematic division, depending upon pecularity of structure and habit, is called a class. There is the class, for example, known as crustaceans, creatures having a crust, which live in the water and breathe through gills. The crab, the crawfish, and the lobster are good examples of these crustaceans. The manner in which they breathe distinguishes them from all other jointed animals. None of the others, though they may live in water, have true, fishlike gills. The crustaceans differ from the other jointed creatures, also, in that they have two pairs of antennæ and varying numbers of legs, always more than four pairs.

This branch of the animal kingdom, the arthropods, boasts another class of creature which is known as the arachnida (this word means "spider family")—the chief representatives of which are the scorpions and spiders. The spiders have four pairs of legs and no antennæ.

Then there is a final small class of these jointed creatures which the scientists call myriapoda, which means "many feet," and of which the centipede is the best known example.

And there are the insects.

So does the outline take further form:

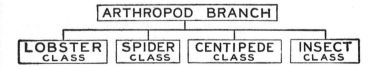

In the orderly arrangement of living creatures we have now arrived at the particular group in which we are especially interested. We proceed to set down certain points that are peculiar to all insects. We find, for instance, that every insect in the world has six legs. No creature that has any other number of legs is an insect. The scientists call them *Hexapods*, which means six-footed.

It is quite a habit to speak of spiders as insects and, in fact, they are popularly considered as belonging to this class of living creatures. Since the insects are so multitudinous in number, since their structure is so well defined, since all the other creatures in the world that are grouped with them hold so strictly to the basic structure common to insects, it becomes necessary to draw the line at the spider, to say to it that it fails in so many respects to live up to the definition of an insect that it must be excluded from that classification. The spider is not an insect.

There is the primary structural distinction of the insects, for instance, that they are creatures with six legs. The spider has eight. There is the secondary structural peculiarity of insects, the fact that the bodies of all of them are divided into three sections, the head, the thorax or chest, and the abdomen. The spider is not so divided; the head and the thorax are not separated, but are joined together in one section of its body, its abdomen making another section. So the spider has but two parts to its body. Thus does it again fail to meet the definition of insect. All insects, further, have antennæ, feelers. The spider has none. It is not an insect.

So is the insect a six-legged animal with antennæ whose body is divided into three sections. Being an arthropod, it is a jointed animal with its skeleton on the outside; it is, in fact, an animal with a jointed body and jointed legs.

The insect and its fellow arthropods have another outstanding peculiarity which make them different from the other members of the animal kingdom. While the blood of the others is red, that of the insects and other arthropods is without color.

One thing that should be borne clearly in mind is the fact that every creature which is a member of the animal kingdom is an animal. In the popular mind it may not be definitely established that man is an animal, that a bird is an animal, that a fish is an animal, that a snail is an animal, that even the tiniest insect is an animal.

Now of all the animals that exist in all the world, there are more different kinds of them that are insects, that have six legs, than there are belonging to all the other classes of animals combined. Some authorities hold that four-fifths of all the creatures of the animal kingdom are six-legged, are insects. There are probably some 250,000 different kinds of insects in the world.

In the task of classification, of sorting, of pigeon-holing these creatures, it becomes necessary to bring them down step by step, always narrowing the field, always tending toward the point where there is before us but the one classified individual for consideration. In the case of the insect the journey of classification from all things to the individual is about half made when all else has been eliminated and but these six-legged creatures are under consideration. In examining insects certain differences immediately begin to assert themselves.

The first subdivision of the six legs is known, by general agreement among the scientists, as an order. In the realm of insects there are in all eighteen orders, but of these six are outstanding, including most of the insects which bear an important relationship to man. Thus may an additional diagram be drawn showing these six orders branching off from the insect class:

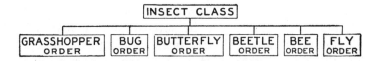

Orders are divided into families. In the world of trees, for instance, the cone-bearing order might be found dividing itself into the pine family, the cypress family, and the yew family. Likewise in the animal world, the order of the birds of prey might be found dividing itself into the owl, the hawk, and the vulture families. So with the flies, which the scientists call *Diptera*, two-wings, we find that they divide themselves into different families, using only three of which we might make the following illustrative diagram:

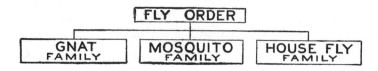

Selecting from this group the mosquitoes as an individual family for further examination, it may be found that they again divide themselves into genera, each genus having certain peculiarities. In the mosquito family there are three well-known genera, indicated as follows:

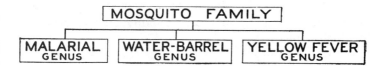

The genus is again divided into species. In the higher animals, for example, cats, in the broad sense of the word, make up a genus. House cats, wild cats, tiger, leopard, lions, are all cats. They are built on the same plan. They are of the same genus. They are known in a "generic" sense as cats. But the house cat, the wild cat, the tiger, the leopard, each considered separately, is a species. Tiger, for instance, is a "specific" name. So is malarial mosquito a generic name because there are a number of species that go to make up the genus, malarial mosquito. We draw a diagram giving the division of this genus into species:

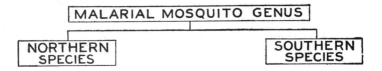

Species divide themselves only into individuals. Thus, finally, beginning with all those elements which go to make up the entire world, we have, through classification, followed an unbroken chain which leads to the individual mosquito which forages its dinner on a summer evening from the back of your hand. Thus may the relationship of that mosquito to the great world be shown. By a similar process may the relationship of any creature to all other creatures be determined. Each may be traced through the kingdom, class, order, family, genus and specie to which it belongs.

So, by retracing this journey, by beginning with the individual mosquito and ending with the world, the structure may be brought to assume its logical form. Such a structure may be built for any living thing. It would look something like this:

```
                         MOSQUITO
                       INDIVIDUAL

                  NORTHERN      SOUTHERN
                  MALARIAL       MALARIAL
                        SPECIES

          MALARIAL        BARREL    YELLOW FEVER
          MOSQUITO       MOSQUITO     MOSQUITO
                        GENUSES

             GNATS     MOSQUITOES    FLIES
                        FAMILIES

  ORTHOPTERA    HEMIPTERA      LEPIDOPTERA   COLEOPTERA
       HYMENOPTERA                      DIPTERA
                        ORDERS

       VERTEBRATES    MOLLUSKS    ARTHROPODS
                        CLASSES

     VEGETABLE                          ANIMAL
                       KINGDOMS

  ANIMATE                              INANIMATE
                      THE WORLD
```

This classification, which the scientists have worked out, of all there is in this world, into

Kingdoms,
> Classes,
>> Orders,
>>> Families,
>>>> Genera,
>>>>> Species,
>>>>>> Individuals,

is the backbone of an orderly study of insects. For generations there have been occasional scientists who have devoted themselves to such investigations, generally as lovers of pure science or as students of nature.

Then, gradually, half a century ago, it began to be realized that there was a practical end to the study of insects. Grasshoppers had swept over Kansas, doing vast damage. Where did they come from and where did they go? Might not a knowledge of the facts prevent them from coming again?

The government became interested. The sum of $5,000 a year was appropriated and three men were set to work on this problem. Thus did this work of the government take form, finally becoming the Bureau of Entomology, of the Department of Agriculture, the biggest bureau for the study of insects in all the world, employing more than 400 scientifically trained men, operating 75 stations, expending nearly $2,500,000 a year.

In the beginning there were almost no trained men. There were no schools in which they might be trained. There was little inducement to take up the study of insects. There were no calls for men who knew about them. The entomologist, invariably pictured with his butterfly net, was, in fact, an object of unending ridicule. The profession had no standing.

Gradually a change has come about. The study of insects has solved such riddles as that of the transmission of yellow fever, of typhoid fever, of bubonic plague, has made it possible to grapple with and defeat these dread diseases. It has made scientific investigation of the insects which cause an annual damage to crops amounting to two billions of dollars, and has steadily lessened that damage. One year it saved the oranges of the nation; another it made it a gift of the Smyrna fig. It has thrown up a barrier about a nation to keep the dangerous insects of distant lands from getting in as did the boll weevil and the gipsy moth. A once mocked profession has become one of dignity and first rank. No state is without its official entomologists. Many counties and cities have them.

No university is without its courses in entomology. Trained and practical men are available by the thousands, economic entomologists who create new wealth by producing better crops, by maintaining a better state of national healthfulness. In a generation the world has been given a practical and applied science where before there had existed only vague theory. This science promises to do much toward making it a better place in which to live.

QUESTIONS

1. (a) In the world of nature one frequently observes the truth of the statement, "Order is Heaven's first law." Explain how the habit of orderly thinking and classifying is invaluable to the student who would make progress in his studies.

(b) Illustrate from the various fields of organized knowledge how classification has helped the student to an understanding of the subject.

2. (a) What specific abilities distinguish animals from plants?

    (b) Briefly indicate in diagram form the big classifications indicated by the author. Be able to explain the diagram and justify your classification.

3. (a) Why do we call the insect and snail as well as the bird and man animals?

    (b) Compare the vertebrates, creatures having backbones, with the arthropods or jointed animals.

    (c) What is the advantage of a backbone to an animal? Of a skeleton inside the body instead of outside?

    (d) What animals carry their houses on their backs?

4. (a) With what sort of covering is an insect provided?

    (b) Why is the spider not an insect?

    (c) Insects outnumber all the other species of animals on the face of the earth. When you read the following stories you will realize how insects hold their own in the air, in the water, and on the land. List all the insects you know according to their families.

    (d) From your general knowledge classify them again as man's helpers or enemies.

5. (a) Economically, insects may furnish or destroy millions of dollars' worth of produce either in the form of fruits or vegetables. What activity has the United States Government engaged in to protect the farmer and increase the wealth of the nation?

    (b) Boys looking forward to a future career may consider the field of entomology as a life work. What opportunities are offered to the progressive student in this important field of work and service?

INDEX

A CATALOGUE OF SELECTED DOVER BOOKS
IN ALL FIELDS OF INTEREST

A CATALOGUE OF SELECTED DOVER BOOKS
IN ALL FIELDS OF INTEREST

WHAT IS SCIENCE?, *N. Campbell*
The role of experiment and measurement, the function of mathematics, the nature of scientific laws, the difference between laws and theories, the limitations of science, and many similarly provocative topics are treated clearly and without technicalities by an eminent scientist. "Still an excellent introduction to scientific philosophy," H. Margenau in *Physics Today*. "A first-rate primer . . . deserves a wide audience," *Scientific American*. 192pp. 5⅜ x 8.
$$S43 \qquad \text{Paperbound } \$1.25$$

THE NATURE OF LIGHT AND COLOUR IN THE OPEN AIR, *M. Minnaert*
Why are shadows sometimes blue, sometimes green, or other colors depending on the light and surroundings? What causes mirages? Why do multiple suns and moons appear in the sky? Professor Minnaert explains these unusual phenomena and hundreds of others in simple, easy-to-understand terms based on optical laws and the properties of light and color. No mathematics is required but artists, scientists, students, and everyone fascinated by these "tricks" of nature will find thousands of useful and amazing pieces of information. Hundreds of observational experiments are suggested which require no special equipment. 200 illustrations; 42 photos. xvi + 362pp. 5⅜ x 8.
$$T196 \qquad \text{Paperbound } \$2.00$$

THE STRANGE STORY OF THE QUANTUM, AN ACCOUNT FOR THE GENERAL READER OF THE GROWTH OF IDEAS UNDERLYING OUR PRESENT ATOMIC KNOWLEDGE, *B. Hoffmann*
Presents lucidly and expertly, with barest amount of mathematics, the problems and theories which led to modern quantum physics. Dr. Hoffmann begins with the closing years of the 19th century, when certain trifling discrepancies were noticed, and with illuminating analogies and examples takes you through the brilliant concepts of Planck, Einstein, Pauli, Broglie, Bohr, Schroedinger, Heisenberg, Dirac, Sommerfeld, Feynman, etc. This edition includes a new, long postscript carrying the story through 1958. "Of the books attempting an account of the history and contents of our modern atomic physics which have come to my attention, this is the best," H. Margenau, Yale University, in *American Journal of Physics*. 32 tables and line illustrations. Index. 275pp. 5⅜ x 8.
$$T518 \qquad \text{Paperbound } \$2.00$$

GREAT IDEAS OF MODERN MATHEMATICS: THEIR NATURE AND USE, *Jagjit Singh*
Reader with only high school math will understand main mathematical ideas of modern physics, astronomy, genetics, psychology, evolution, etc. better than many who use them as tools, but comprehend little of their basic structure. Author uses his wide knowledge of non-mathematical fields in brilliant exposition of differential equations, matrices, group theory, logic, statistics, problems of mathematical foundations, imaginary numbers, vectors, etc. Original publication. 2 appendixes. 2 indexes. 65 ills. 322pp. 5⅜ x 8.
$$T587 \qquad \text{Paperbound } \$2.00$$

A Short Account of the History of Mathematics,
W. W. Rouse Ball
Last previous edition (1908) hailed by mathematicians and laymen for lucid overview of math as living science, for understandable presentation of individual contributions of great mathematicians. Treats lives, discoveries of every important school and figure from Egypt, Phoenicia to late nineteenth century. Greek schools of Ionia, Cyzicus, Alexandria, Byzantium, Pythagoras; primitive arithmetic; Middle Ages and Renaissance, including European and Asiatic contributions; modern math of Descartes, Pascal, Wallis, Huygens, Newton, Euler, Lambert, Laplace, scores more. More emphasis on historical development, exposition of ideas than other books on subject. Non-technical, readable text can be followed with no more preparation than high-school algebra. Index. 544pp. 5⅜ x 8. Paperbound $2.25

Great Ideas and Theories of Modern Cosmology, *Jagjit Singh*
Companion volume to author's popular "Great Ideas of Modern Mathematics" (Dover, $2.00). The best non-technical survey of post-Einstein attempts to answer perhaps unanswerable questions of origin, age of Universe, possibility of life on other worlds, etc. Fundamental theories of cosmology and cosmogony recounted, explained, evaluated in light of most recent data: Einstein's concepts of relativity, space-time; Milne's a priori world-system; astrophysical theories of Jeans, Eddington; Hoyle's "continuous creation;" contributions of dozens more scientists. A faithful, comprehensive critical summary of complex material presented in an extremely well-written text intended for laymen. Original publication. Index. xii + 276pp. 5⅜ x 8½. Paperbound $2.00

The Restless Universe, *Max Born*
A remarkably lucid account by a Nobel Laureate of recent theories of wave mechanics, behavior of gases, electrons and ions, waves and particles, electronic structure of the atom, nuclear physics, and similar topics. "Much more thorough and deeper than most attempts . . . easy and delightful," *Chemical and Engineering News*. Special feature: 7 animated sequences of 60 figures each showing such phenomena as gas molecules in motion, the scattering of alpha particles, etc. 11 full-page plates of photographs. Total of nearly 600 illustrations. 351pp. 6⅛ x 9¼. Paperbound $2.00

Planets, Stars and Galaxies: Descriptive Astronomy for Beginners,
A. E. Fanning
What causes the progression of the seasons? Phases of the moon? The Aurora Borealis? How much does the sun weigh? What are the chances of life on our sister planets? Absorbing introduction to astronomy, incorporating the latest discoveries and theories: the solar wind, the surface temperature of Venus, the pock-marked face of Mars, quasars, and much more. Places you on the frontiers of one of the most vital sciences of our time. Revised (1966). Introduction by Donald H. Menzel, Harvard University. References. Index. 45 illustrations. 189pp. 5¼ x 8¼. Paperbound $1.50

Great Ideas in Information Theory, Language and Cybernetics,
Jagjit Singh
Non-mathematical, but profound study of information, language, the codes used by men and machines to communicate, the principles of analog and digital computers, work of McCulloch, Pitts, von Neumann, Turing, and Uttley, correspondences between intricate mechanical network of "thinking machines" and more intricate neurophysiological mechanism of human brain. Indexes. 118 figures. 50 tables. ix + 338pp. 5⅜ x 8½. Paperbound $2.00

THE MUSIC OF THE SPHERES: THE MATERIAL UNIVERSE—FROM ATOM TO QUASAR, SIMPLY EXPLAINED, *Guy Murchie*
Vast compendium of fact, modern concept and theory, observed and calculated data, historical background guides intelligent layman through the material universe. Brilliant exposition of earth's construction, explanations for moon's craters, atmospheric components of Venus and Mars (with data from recent fly-by's), sun spots, sequences of star birth and death, neighboring galaxies, contributions of Galileo, Tycho Brahe, Kepler, etc.; and (Vol. 2) construction of the atom (describing newly discovered sigma and xi subatomic particles), theories of sound, color and light, space and time, including relativity theory, quantum theory, wave theory, probability theory, work of Newton, Maxwell, Faraday, Einstein, de Broglie, etc. "Best presentation yet offered to the intelligent general reader," *Saturday Review*. Revised (1967). Index. 319 illustrations by the author. Total of xx + 644pp. 5⅜ x 8½.

Vol. 1 Paperbound $2.00, Vol. 2 Paperbound $2.00,
The set $4.00

FOUR LECTURES ON RELATIVITY AND SPACE, *Charles Proteus Steinmetz*
Lecture series, given by great mathematician and electrical engineer, generally considered one of the best popular-level expositions of special and general relativity theories and related questions. Steinmetz translates complex mathematical reasoning into language accessible to laymen through analogy, example and comparison. Among topics covered are relativity of motion, location, time; of mass; acceleration; 4-dimensional time-space; geometry of the gravitational field; curvature and bending of space; non-Euclidean geometry. Index. 40 illustrations. x + 142pp. 5⅜ x 8½.   Paperbound $1.35

HOW TO KNOW THE WILD FLOWERS, *Mrs. William Starr Dana*
Classic nature book that has introduced thousands to wonders of American wild flowers. Color-season principle of organization is easy to use, even by those with no botanical training, and the genial, refreshing discussions of history, folklore, uses of over 1,000 native and escape flowers, foliage plants are informative as well as fun to read. Over 170 full-page plates, collected from several editions, may be colored in to make permanent records of finds. Revised to conform with 1950 edition of Gray's Manual of Botany. xlii + 438pp. 5⅜ x 8½.   Paperbound $2.00

MANUAL OF THE TREES OF NORTH AMERICA, *Charles Sprague Sargent*
Still unsurpassed as most comprehensive, reliable study of North American tree characteristics, precise locations and distribution. By dean of American dendrologists. Every tree native to U.S., Canada, Alaska; 185 genera, 717 species, described in detail—leaves, flowers, fruit, winterbuds, bark, wood, growth habits, etc. plus discussion of varieties and local variants, immaturity variations. Over 100 keys, including unusual 11-page analytical key to genera, aid in identification. 783 clear illustrations of flowers, fruit, leaves. An unmatched permanent reference work for all nature lovers. Second enlarged (1926) edition. Synopsis of families. Analytical key to genera. Glossary of technical terms. Index. 783 illustrations, 1 map. Total of 982pp. 5⅜ x 8.

Vol. 1 Paperbound $2.25, Vol. 2 Paperbound $2.25,
The set $4.50

It's Fun to Make Things From Scrap Materials,
Evelyn Glantz Hershoff
What use are empty spools, tin cans, bottle tops? What can be made from rubber bands, clothes pins, paper clips, and buttons? This book provides simply worded instructions and large diagrams showing you how to make cookie cutters, toy trucks, paper turkeys, Halloween masks, telephone sets, aprons, linoleum block- and spatter prints — in all 399 projects! Many are easy enough for young children to figure out for themselves; some challenging enough to entertain adults; all are remarkably ingenious ways to make things from materials that cost pennies or less! Formerly "Scrap Fun for Everyone." Index. 214 illustrations. 373pp. 5⅜ x 8½.   Paperbound $1.50

Symbolic Logic and The Game of Logic, *Lewis Carroll*
"Symbolic Logic" is not concerned with modern symbolic logic, but is instead a collection of over 380 problems posed with charm and imagination, using the syllogism and a fascinating diagrammatic method of drawing conclusions. In "The Game of Logic" Carroll's whimsical imagination devises a logical game played with 2 diagrams and counters (included) to manipulate hundreds of tricky syllogisms. The final section, "Hit or Miss" is a lagniappe of 101 additional puzzles in the delightful Carroll manner. Until this reprint edition, both of these books were rarities costing up to $15 each. Symbolic Logic: Index. xxxi + 199pp. The Game of Logic: 96pp. 2 vols. bound as one. 5⅜ x 8.
Paperbound $2.00

Mathematical Puzzles of Sam Loyd, Part I
selected and edited by M. Gardner
Choice puzzles by the greatest American puzzle creator and innovator. Selected from his famous collection, "Cyclopedia of Puzzles," they retain the unique style and historical flavor of the originals. There are posers based on arithmetic, algebra, probability, game theory, route tracing, topology, counter and sliding block, operations research, geometrical dissection. Includes the famous "14-15" puzzle which was a national craze, and his "Horse of a Different Color" which sold millions of copies. 117 of his most ingenious puzzles in all. 120 line drawings and diagrams. Solutions. Selected references. xx + 167pp. 5⅜ x 8.
Paperbound $1.00

String Figures and How to Make Them, *Caroline Furness Jayne*
107 string figures plus variations selected from the best primitive and modern examples developed by Navajo, Apache, pygmies of Africa, Eskimo, in Europe, Australia, China, etc. The most readily understandable, easy-to-follow book in English on perennially popular recreation. Crystal-clear exposition; step-by-step diagrams. Everyone from kindergarten children to adults looking for unusual diversion will be endlessly amused. Index. Bibliography. Introduction by A. C. Haddon. 17 full-page plates, 960 illustrations. xxiii + 401pp. 5⅜ x 8½.
Paperbound $2.00

Paper Folding for Beginners, *W. D. Murray and F. J. Rigney*
A delightful introduction to the varied and entertaining Japanese art of origami (paper folding), with a full, crystal-clear text that anticipates every difficulty; over 275 clearly labeled diagrams of all important stages in creation. You get results at each stage, since complex figures are logically developed from simpler ones. 43 different pieces are explained: sailboats, frogs, roosters, etc. 6 photographic plates. 279 diagrams. 95pp. 5⅜ x 8⅜. Paperbound $1.00

PRINCIPLES OF ART HISTORY,
H. Wölfflin

Analyzing such terms as "baroque," "classic," "neoclassic," "primitive," "picturesque," and 164 different works by artists like Botticelli, van Cleve, Dürer, Hobbema, Holbein, Hals, Rembrandt, Titian, Brueghel, Vermeer, and many others, the author establishes the classifications of art history and style on a firm, concrete basis. This classic of art criticism shows what really occurred between the 14th-century primitives and the sophistication of the 18th century in terms of basic attitudes and philosophies. "A remarkable lesson in the art of seeing," *Sat. Rev. of Literature*. Translated from the 7th German edition. 150 illustrations. 254pp. 6⅛ x 9¼. Paperbound $2.00

PRIMITIVE ART,
Franz Boas

This authoritative and exhaustive work by a great American anthropologist covers the entire gamut of primitive art. Pottery, leatherwork, metal work, stone work, wood, basketry, are treated in detail. Theories of primitive art, historical depth in art history, technical virtuosity, unconscious levels of patterning, symbolism, styles, literature, music, dance, etc. A must book for the interested layman, the anthropologist, artist, handicrafter (hundreds of unusual motifs), and the historian. Over 900 illustrations (50 ceramic vessels, 12 totem poles, etc.). 376pp. 5⅜ x 8. Paperbound $2.25

THE GENTLEMAN AND CABINET MAKER'S DIRECTOR,
Thomas Chippendale

A reprint of the 1762 catalogue of furniture designs that went on to influence generations of English and Colonial and Early Republic American furniture makers. The 200 plates, most of them full-page sized, show Chippendale's designs for French (Louis XV), Gothic, and Chinese-manner chairs, sofas, canopy and dome beds, cornices, chamber organs, cabinets, shaving tables, commodes, picture frames, frets, candle stands, chimney pieces, decorations, etc. The drawings are all elegant and highly detailed; many include construction diagrams and elevations. A supplement of 24 photographs shows surviving pieces of original and Chippendale-style pieces of furniture. Brief biography of Chippendale by N. I. Bienenstock, editor of *Furniture World*. Reproduced from the 1762 edition. 200 plates, plus 19 photographic plates. vi + 249pp. 9⅛ x 12¼. Paperbound $3.50

AMERICAN ANTIQUE FURNITURE: A BOOK FOR AMATEURS,
Edgar G. Miller, Jr.

Standard introduction and practical guide to identification of valuable American antique furniture. 2115 illustrations, mostly photographs taken by the author in 148 private homes, are arranged in chronological order in extensive chapters on chairs, sofas, chests, desks, bedsteads, mirrors, tables, clocks, and other articles. Focus is on furniture accessible to the collector, including simpler pieces and a larger than usual coverage of Empire style. Introductory chapters identify structural elements, characteristics of various styles, how to avoid fakes, etc. "We are frequently asked to name some book on American furniture that will meet the requirements of the novice collector, the beginning dealer, and . . . the general public. . . . We believe Mr. Miller's two volumes more completely satisfy this specification than any other work," *Antiques*. Appendix. Index. Total of vi + 1106pp. 7⅞ x 10¾.

Two volume set, paperbound $7.50

THE BAD CHILD'S BOOK OF BEASTS, MORE BEASTS FOR WORSE CHILDREN, and A MORAL ALPHABET, *H. Belloc*

Hardly and anthology of humorous verse has appeared in the last 50 years without at least a couple of these famous nonsense verses. But one must see the entire volumes — with all the delightful original illustrations by Sir Basil Blackwood — to appreciate fully Belloc's charming and witty verses that play so subacidly on the platitudes of life and morals that beset his day — and ours. A great humor classic. Three books in one. Total of 157pp. 5⅜ x 8.

Paperbound $1.00

THE DEVIL'S DICTIONARY, *Ambrose Bierce*

Sardonic and irreverent barbs puncturing the pomposities and absurdities of American politics, business, religion, literature, and arts, by the country's greatest satirist in the classic tradition. Epigrammatic as Shaw, piercing as Swift, American as Mark Twain, Will Rogers, and Fred Allen, Bierce will always remain the favorite of a small coterie of enthusiasts, and of writers and speakers whom he supplies with "some of the most gorgeous witticisms of the English language" (H. L. Mencken). Over 1000 entries in alphabetical order. 144pp. 5⅜ x 8. Paperbound $1.00

THE COMPLETE NONSENSE OF EDWARD LEAR.

This is the only complete edition of this master of gentle madness available at a popular price. *A Book of Nonsense, Nonsense Songs, More Nonsense Songs and Stories* in their entirety with all the old favorites that have delighted children and adults for years. The Dong With A Luminous Nose, The Jumblies, The Owl and the Pussycat, and hundreds of other bits of wonderful nonsense. 214 limericks, 3 sets of Nonsense Botany, 5 Nonsense Alphabets, 546 drawings by Lear himself, and much more. 320pp. 5⅜ x 8. Paperbound $1.00

THE WIT AND HUMOR OF OSCAR WILDE, *ed. by Alvin Redman*

Wilde at his most brilliant, in 1000 epigrams exposing weaknesses and hypocrisies of "civilized" society. Divided into 49 categories—sin, wealth, women, America, etc.—to aid writers, speakers. Includes excerpts from his trials, books, plays, criticism. Formerly "The Epigrams of Oscar Wilde." Introduction by Vyvyan Holland, Wilde's only living son. Introductory essay by editor. 260pp. 5⅜ x 8. Paperbound $1.00

A CHILD'S PRIMER OF NATURAL HISTORY, *Oliver Herford*

Scarcely an anthology of whimsy and humor has appeared in the last 50 years without a contribution from Oliver Herford. Yet the works from which these examples are drawn have been almost impossible to obtain! Here at last are Herford's improbable definitions of a menagerie of familiar and weird animals, each verse illustrated by the author's own drawings. 24 drawings in 2 colors; 24 additional drawings. vii + 95pp. 6½ x 6. Paperbound $1.00

THE BROWNIES: THEIR BOOK, *Palmer Cox*

The book that made the Brownies a household word. Generations of readers have enjoyed the antics, predicaments and adventures of these jovial sprites, who emerge from the forest at night to play or to come to the aid of a deserving human. Delightful illustrations by the author decorate nearly every page. 24 short verse tales with 266 illustrations. 155pp. 6⅝ x 9¼.

Paperbound $1.50

THE PRINCIPLES OF PSYCHOLOGY,
William James

The full long-course, unabridged, of one of the great classics of Western literature and science. Wonderfully lucid descriptions of human mental activity, the stream of thought, consciousness, time perception, memory, imagination, emotions, reason, abnormal phenomena, and similar topics. Original contributions are integrated with the work of such men as Berkeley, Binet, Mills, Darwin, Hume, Kant, Royce, Schopenhauer, Spinoza, Locke, Descartes, Galton, Wundt, Lotze, Herbart, Fechner, and scores of others. All contrasting interpretations of mental phenomena are examined in detail—introspective analysis, philosophical interpretation, and experimental research. "A classic," *Journal of Consulting Psychology.* "The main lines are as valid as ever," *Psychoanalytical Quarterly.* "Standard reading ... a classic of interpretation," *Psychiatric Quarterly.* 94 illustrations. 1408pp. 5⅜ x 8.

Vol. 1 Paperbound $2.50, Vol. 2 Paperbound $2.50,
The set $5.00

VISUAL ILLUSIONS: THEIR CAUSES, CHARACTERISTICS AND APPLICATIONS,
M. Luckiesh

"Seeing is deceiving," asserts the author of this introduction to virtually every type of optical illusion known. The text both describes and explains the principles involved in color illusions, figure-ground, distance illusions, etc. 100 photographs, drawings and diagrams prove how easy it is to fool the sense: circles that aren't round, parallel lines that seem to bend, stationary figures that seem to move as you stare at them — illustration after illustration strains our credulity at what we see. Fascinating book from many points of view, from applications for artists, in camouflage, etc. to the psychology of vision. New introduction by William Ittleson, Dept. of Psychology, Queens College. Index. Bibliography. xxi + 252pp. 5⅜ x 8½. Paperbound $1.50

FADS AND FALLACIES IN THE NAME OF SCIENCE,
Martin Gardner

This is the standard account of various cults, quack systems, and delusions which have masqueraded as science: hollow earth fanatics. Reich and orgone sex energy, dianetics, Atlantis, multiple moons, Forteanism, flying saucers, medical fallacies like iridiagnosis, zone therapy, etc. A new chapter has been added on Bridey Murphy, psionics, and other recent manifestations in this field. This is a fair, reasoned appraisal of eccentric theory which provides excellent inoculation against cleverly masked nonsense. "Should be read by everyone, scientist and non-scientist alike," R. T. Birge, Prof. Emeritus of Physics, Univ. of California; Former President, American Physical Society. Index. x + 365pp. 5⅜ x 8. Paperbound $1.85

ILLUSIONS AND DELUSIONS OF THE SUPERNATURAL AND THE OCCULT,
D. H. Rawcliffe

Holds up to rational examination hundreds of persistent delusions including crystal gazing, automatic writing, table turning, mediumistic trances, mental healing, stigmata, lycanthropy, live burial, the Indian Rope Trick, spiritualism, dowsing, telepathy, clairvoyance, ghosts, ESP, etc. The author explains and exposes the mental and physical deceptions involved, making this not only an exposé of supernatural phenomena, but a valuable exposition of characteristic types of abnormal psychology. Originally titled "The Psychology of the Occult." 14 illustrations. Index. 551pp. 5⅜ x 8. Paperbound $2.25

FAIRY TALE COLLECTIONS, *edited by Andrew Lang*
Andrew Lang's fairy tale collections make up the richest shelf-full of traditional children's stories anywhere available. Lang supervised the translation of stories from all over the world—familiar European tales collected by Grimm, animal stories from Negro Africa, myths of primitive Australia, stories from Russia, Hungary, Iceland, Japan, and many other countries. Lang's selection of translations are unusually high; many authorities consider that the most familiar tales find their best versions in these volumes. All collections are richly decorated and illustrated by H. J. Ford and other artists.

THE BLUE FAIRY BOOK. 37 stories. 138 illustrations. ix + 390pp. 5⅜ x 8½.
Paperbound $1.50

THE GREEN FAIRY BOOK. 42 stories. 100 illustrations. xiii + 366pp. 5⅜ x 8½.
Paperbound $1.50

THE BROWN FAIRY BOOK. 32 stories. 50 illustrations, 8 in color. xii + 350pp. 5⅜ x 8½.
Paperbound $1.50

THE BEST TALES OF HOFFMANN, *edited by E. F. Bleiler*
10 stories by E. T. A. Hoffmann, one of the greatest of all writers of fantasy. The tales include "The Golden Flower Pot," "Automata," "A New Year's Eve Adventure," "Nutcracker and the King of Mice," "Sand-Man," and others. Vigorous characterizations of highly eccentric personalities, remarkably imaginative situations, and intensely fast pacing has made these tales popular all over the world for 150 years. Editor's introduction. 7 drawings by Hoffmann. xxxiii + 419pp. 5⅜ x 8½.
Paperbound $2.00

GHOST AND HORROR STORIES OF AMBROSE BIERCE, *edited by E. F. Bleiler*
Morbid, eerie, horrifying tales of possessed poets, shabby aristocrats, revived corpses, and haunted malefactors. Widely acknowledged as the best of their kind between Poe and the moderns, reflecting their author's inner torment and bitter view of life. Includes "Damned Thing," "The Middle Toe of the Right Foot," "The Eyes of the Panther," "Visions of the Night," "Moxon's Master," and over a dozen others. Editor's introduction. xxii + 199pp. 5⅜ x 8½.
Paperbound $1.25

THREE GOTHIC NOVELS, *edited by E. F. Bleiler*
Originators of the still popular Gothic novel form, influential in ushering in early 19th-century Romanticism. Horace Walpole's *Castle of Otranto*, William Beckford's *Vathek*, John Polidori's *The Vampyre*, and a *Fragment* by Lord Byron are enjoyable as exciting reading or as documents in the history of English literature. Editor's introduction. xi + 291pp. 5⅜ x 8½.
Paperbound $2.00

BEST GHOST STORIES OF LEFANU, *edited by E. F. Bleiler*
Though admired by such critics as V. S. Pritchett, Charles Dickens and Henry James, ghost stories by the Irish novelist Joseph Sheridan LeFanu have never become as widely known as his detective fiction. About half of the 16 stories in this collection have never before been available in America. Collection includes "Carmilla" (perhaps the best vampire story ever written), "The Haunted Baronet," "The Fortunes of Sir Robert Ardagh," and the classic "Green Tea." Editor's introduction. 7 contemporary illustrations. Portrait of LeFanu. xii + 467pp. 5⅜ x 8.
Paperbound $2.00

EASY-TO-DO ENTERTAINMENTS AND DIVERSIONS WITH COINS, CARDS, STRING, PAPER AND MATCHES, *R. M. Abraham*

Over 300 tricks, games and puzzles will provide young readers with absorbing fun. Sections on card games; paper-folding; tricks with coins, matches and pieces of string; games for the agile; toy-making from common household objects; mathematical recreations; and 50 miscellaneous pastimes. Anyone in charge of groups of youngsters, including hard-pressed parents, and in need of suggestions on how to keep children sensibly amused and quietly content will find this book indispensable. Clear, simple text, copious number of delightful line drawings and illustrative diagrams. Originally titled "Winter Nights' Entertainments." Introduction by Lord Baden Powell. 329 illustrations. v + 186pp. 5⅜ x 8½. Paperbound $1.00

AN INTRODUCTION TO CHESS MOVES AND TACTICS SIMPLY EXPLAINED, *Leonard Barden*

Beginner's introduction to the royal game. Names, possible moves of the pieces, definitions of essential terms, how games are won, etc. explained in 30-odd pages. With this background you'll be able to sit right down and play. Balance of book teaches strategy — openings, middle game, typical endgame play, and suggestions for improving your game. A sample game is fully analyzed. True middle-level introduction, teaching you all the essentials without oversimplifying or losing you in a maze of detail. 58 figures. 102pp. 5⅜ x 8½. Paperbound $1.00

LASKER'S MANUAL OF CHESS, *Dr. Emanuel Lasker*

Probably the greatest chess player of modern times, Dr. Emanuel Lasker held the world championship 28 years, independent of passing schools or fashions. This unmatched study of the game, chiefly for intermediate to skilled players, analyzes basic methods, combinations, position play, the aesthetics of chess, dozens of different openings, etc., with constant reference to great modern games. Contains a brilliant exposition of Steinitz's important theories. Introduction by Fred Reinfeld. Tables of Lasker's tournament record. 3 indices. 308 diagrams. 1 photograph. xxx + 349pp. 5⅜ x 8. Paperbound $2.25

COMBINATIONS: THE HEART OF CHESS, *Irving Chernev*

Step-by-step from simple combinations to complex, this book, by a well-known chess writer, shows you the intricacies of pins, counter-pins, knight forks, and smothered mates. Other chapters show alternate lines of play to those taken in actual championship games; boomerang combinations; classic examples of brilliant combination play by Nimzovich, Rubinstein, Tarrasch, Botvinnik, Alekhine and Capablanca. Index. 356 diagrams. ix + 245pp. 5⅜ x 8½. Paperbound $1.85

HOW TO SOLVE CHESS PROBLEMS, *K. S. Howard*

Full of practical suggestions for the fan or the beginner — who knows only the moves of the chessmen. Contains preliminary section and 58 two-move, 46 three-move, and 8 four-move problems composed by 27 outstanding American problem creators in the last 30 years. Explanation of all terms and exhaustive index. "Just what is wanted for the student," Brian Harley. 112 problems, solutions. vi + 171pp. 5⅜ x 8. Paperbound $1.35

SOCIAL THOUGHT FROM LORE TO SCIENCE,
H. E. Barnes and H. Becker
An immense survey of sociological thought and ways of viewing, studying, planning, and reforming society from earliest times to the present. Includes thought on society of preliterate peoples, ancient non-Western cultures, and every great movement in Europe, America, and modern Japan. Analyzes hundreds of great thinkers: Plato, Augustine, Bodin, Vico, Montesquieu, Herder, Comte, Marx, etc. Weighs the contributions of utopians, sophists, fascists and communists; economists, jurists, philosophers, ecclesiastics, and every 19th and 20th century school of scientific sociology, anthropology, and social psychology throughout the world. Combines topical, chronological, and regional approaches, treating the evolution of social thought as a process rather than as a series of mere topics. "Impressive accuracy, competence, and discrimination . . . easily the best single survey," *Nation*. Thoroughly revised, with new material up to 1960. 2 indexes. Over 2200 bibliographical notes. Three volume set. Total of 1586pp. 5⅜ x 8.
Vol. 1 Paperbound $2.75, Vol. 2 Paperbound $2.75, Vol. 3 Paperbound $2.50
The set $8.00

A HISTORY OF HISTORICAL WRITING, *Harry Elmer Barnes*
Virtually the only adequate survey of the whole course of historical writing in a single volume. Surveys developments from the beginnings of historiography in the ancient Near East and the Classical World, up through the Cold War. Covers major historians in detail, shows interrelationship with cultural background, makes clear individual contributions, evaluates and estimates importance; also enormously rich upon minor authors and thinkers who are usually passed over. Packed with scholarship and learning, clear, easily written. Indispensable to every student of history. Revised and enlarged up to 1961. Index and bibliography. xv + 442pp. 5⅜ x 8½.   Paperbound $2.50

JOHANN SEBASTIAN BACH, *Philipp Spitta*
The complete and unabridged text of the definitive study of Bach. Written some 70 years ago, it is still unsurpassed for its coverage of nearly all aspects of Bach's life and work. There could hardly be a finer non-technical introduction to Bach's music than the detailed, lucid analyses which Spitta provides for hundreds of individual pieces. 26 solid pages are devoted to the B minor mass, for example, and 30 pages to the glorious St. Matthew Passion. This monumental set also includes a major analysis of the music of the 18th century: Buxtehude, Pachelbel, etc. "Unchallenged as the last word on one of the supreme geniuses of music," John Barkham, *Saturday Review Syndicate*. Total of 1819pp. Heavy cloth binding. 5⅜ x 8.
Two volume set, clothbound $13.50

BEETHOVEN AND HIS NINE SYMPHONIES, *George Grove*
In this modern middle-level classic of musicology Grove not only analyzes all nine of Beethoven's symphonies very thoroughly in terms of their musical structure, but also discusses the circumstances under which they were written, Beethoven's stylistic development, and much other background material. This is an extremely rich book, yet very easily followed; it is highly recommended to anyone seriously interested in music. Over 250 musical passages. Index. viii + 407pp. 5⅜ x 8.   Paperbound $2.00

THREE SCIENCE FICTION NOVELS,
John Taine
Acknowledged by many as the best SF writer of the 1920's, Taine (under the name Eric Temple Bell) was also a Professor of Mathematics of considerable renown. Reprinted here are *The Time Stream*, generally considered Taine's best, *The Greatest Game*, a biological-fiction novel, and *The Purple Sapphire*, involving a supercivilization of the past. Taine's stories tie fantastic narratives to frameworks of original and logical scientific concepts. Speculation is often profound on such questions as the nature of time, concept of entropy, cyclical universes, etc. 4 contemporary illustrations. v + 532pp. 5⅜ x 8⅜.

T1180   Paperbound $2.00

SEVEN SCIENCE FICTION NOVELS,
H. G. Wells
Full unabridged texts of 7 science-fiction novels of the master. Ranging from biology, physics, chemistry, astronomy, to sociology and other studies, Mr. Wells extrapolates whole worlds of strange and intriguing character. "One will have to go far to match this for entertainment, excitement, and sheer pleasure . . ."*New York Times*. Contents: The Time Machine, The Island of Dr. Moreau, The First Men in the Moon, The Invisible Man, The War of the Worlds, The Food of the Gods, In The Days of the Comet. 1015pp. 5⅜ x 8.

T264   Clothbound $5.00

28 SCIENCE FICTION STORIES OF H. G. WELLS.
Two full, unabridged novels, *Men Like Gods* and *Star Begotten*, plus 26 short stories by the master science-fiction writer of all time! Stories of space, time, invention, exploration, futuristic adventure. Partial contents: *The Country of the Blind, In the Abyss, The Crystal Egg, The Man Who Could Work Miracles, A Story of Days to Come, The Empire of the Ants, The Magic Shop, The Valley of the Spiders, A Story of the Stone Age, Under the Knife, Sea Raiders,* etc. An indispensable collection for the library of anyone interested in science fiction adventure. 928pp. 5⅜ x 8.   T265   Clothbound $5.00

THREE MARTIAN NOVELS,
Edgar Rice Burroughs
Complete, unabridged reprinting, in one volume, of Thuvia, Maid of Mars; Chessmen of Mars; The Master Mind of Mars. Hours of science-fiction adventure by a modern master storyteller. Reset in large clear type for easy reading. 16 illustrations by J. Allen St. John. vi + 490pp. 5⅜ x 8½.

T39   Paperbound $2.50

AN INTELLECTUAL AND CULTURAL HISTORY OF THE WESTERN WORLD,
Harry Elmer Barnes
Monumental 3-volume survey of intellectual development of Europe from primitive cultures to the present day. Every significant product of human intellect traced through history: art, literature, mathematics, physical sciences, medicine, music, technology, social sciences, religions, jurisprudence, education, etc. Presentation is lucid and specific, analyzing in detail specific discoveries, theories, literary works, and so on. Revised (1965) by recognized scholars in specialized fields under the direction of Prof. Barnes. Revised bibliography. Indexes. 24 illustrations. Total of xxix + 1318pp.

T1275, T1276, T1277   Three volume set, paperbound $7.50

FABLES OF AESOP,
according to Sir Roger L'Estrange, with 50 drawings by Alexander Calder
Republication of rare 1931 Paris edition (limited to 665 copies) of 200 fables by Aesop in the 1692 L'Estrange translation. Illustrated with 50 highly imaginative, witty and occasionally ribald line drawings by the inventor of "mobiles" and "stabiles." "Fifty wonderfully inventive Alexander Calder drawings, impertinent as any of the artist's wire sculptures, make a delightful, modern counterpoint to the thoroughly moral tales," *Saturday Review.* 124pp. 6½ x 9¼. Paperbound $1.25

DRAWINGS OF REMBRANDT
One of the earliest and best collections of Rembrandt drawings—the Lippmann-Hofstede de Groot facsimiles (1888)—is here reproduced in entirety. Collection contains 550 faithfully reproduced drawings in inks, chalks, and silverpoint; some, hasty sketches recorded on a handy scrap of paper; others, studies for well-known oil paintings. Edited, with scholarly commentary by Seymour Slive, Harvard University. "In large matters of appearance, size (9 x 12-inch page), paper color and weight, uniformity of plate texture, typography and printing, these two volumes could scarcely be improved," *Arts and Architecture.* "Altogether commendable . . . among the year's best," *New York Times.* Editor's introduction, notes. 3 indexes, 2 concordances. Total of lxxix + 552pp. 9⅛ x 12¼. Two volume set, paperbound $6.00
Two volume set, clothbound $12.50

THE EARLY WORK OF AUBREY BEARDSLEY
Together with *The Later Work*, the standard source for the most important Beardsley drawings. Edited by John Lane, *Early Work* contains 157 full-page plates including Burne-Jones style work, the *Morte d'Arthur* series, cover designs and illustrations from *The Studio* and other magazines, theatre posters, "Kiss of Judas," "Seigfried," portraits of himself, Emile Zola, and Verdi, and illustrations for Wilde's play *Salome.* 2 color plates. Introduction by H. C. Marillier. xii + 175pp. 8⅛ x 11. Paperbound $2.50
Clothbound $8.50

THE LATER WORK OF AUBREY BEARDSLEY
Edited by John Lane, collection contains 174 full-page plates including *Savoy* and *Yellow Book* illustrations, book plates, "The Wagnerites," "La Dame aux Camellias," selections from *Lysistrata*, illustrations to *Das Rheingold, Venus and Tannhauser*, and the "Rape of the Lock" series. 2 color plates. xiv + 174pp. 8⅛ x 11. Paperbound $2.50
Clothbound $8.50

Prices subject to change without notice.

Available at your book dealer or write for free catalogue to Dept. Adsci, Dover Publications, Inc., 180 Varick St., N.Y., N.Y. 10014. Dover publishes more than 150 books each year on science, elementary and advanced mathematics, biology, music, art, literary history, social sciences and other areas.

Pan